Cities Research Series

Series Editor

This book series brings together researchers, planning professionals and policy makers in the area of cities and urban development and publishes recent advances in the field. It addresses contemporary urban issues to understand and meet urban challenges and make (future) cities more sustainable and better places to live. The series covers, but is not limited to the following topics:

- Transport policy and behaviour
- Architecture, architectural science and construction engineering
- Urban planning, urban design and housing
- Infrastructure planning and management
- Complex systems and cities
- Urban and regional governance
- Smart and digital technologies

Scott Baum · Emma Baker · Amanda Davies ·
John Stone · Elizabeth Taylor

Pandemic Cities

The COVID-19 Crisis and Australian Urban Regions

Springer

Scott Baum
Griffith University
Brisbane, QLD, Australia

Amanda Davies
School of Social Sciences
University of Western Australia
Perth, WA, Australia

Elizabeth Taylor
Urban Planning and Design
Monash University
Clayton Victoria, VIC, Australia

Emma Baker
School of Social Sciences
University of Adelaide
Adelaide, SA, Australia

John Stone
Architecture, Building and Planning
University of Melbourne
Parkville, VIC, Australia

ISSN 2662-4842 ISSN 2662-4850 (electronic)
Cities Research Series
ISBN 978-981-19-5883-0 ISBN 978-981-19-5884-7 (eBook)
https://doi.org/10.1007/978-981-19-5884-7

This Springer imprint is published by the registered company Springer Nature Singapore Pte Ltd.
The registered company address is: 152 Beach Road, #21-01/04 Gateway East, Singapore 189721, Singapore

Foreword to Pandemic Cities by Baum et al

Viral infections and other transmissible diseases have forever shaped cities, literally and metaphorically. The recent or perhaps still current COVID pandemic prompted many commentators and politicians to speak of its scale and impact as unprecedented. This revealed their lack of historical understanding and exemplified a rather disturbing tendency to neglect important lessons from history.

This latest contribution to the *Cities Research Series* provides a valuable corrective to such a limited historical awareness. As a unique collaboration among some of Australia's leading scholars of city life and urban processes, it explores many of the most important areas impacted by the pandemic, including housing, work, mobility and population change.

It is sometimes said that momentous events like the pandemic act like Barium Meals, revealing features that otherwise are not visible or apparent. While many urban scholars have been cataloguing the inequalities and inequities of urban life for centuries, the last few years have shown us the extent of many urban vulnerabilities that have often been ignored by policymakers and city leaders. We have become more acutely aware of the essential services delivered by some of our lowest paid workers, of the mental stresses that occur when we are unable to socialise, especially in green spaces, and the fragility of many of the supply chains on which we have come to depend. But we have also learnt that when we have to, we as individuals and as members of communities and our governments can choose to do things very differently. We can house street homeless people in inner city hotels, we can provide financial support to everyone without means of testing, we can change the way we work (including academics) very quickly, and we can develop vaccines and other medicines much more quickly than we have done previously. Of course, many of these measures are extremely costly, and building those costs into ongoing budgets might not be feasible. But, we have learnt that these are choices we make rather than impossibilities.

I am delighted that this book, drawing on the collective expertise of five renowned urban scholars in Australia, provides some historical insights, trenchant contemporary analysis of the impact of the pandemic on Australian cities and provocative questions about the future of our cities. While focussed on the Australian experience, the book will be of interest and value to those who live in and study cities around the world.

Prof. Paul Burton
Series Editor
Cities Research Institute
Griffith University
Southport, QLD, Australia
p.burton@griffith.edu.au

Preface

This book highlights the impact of the COVID-19 pandemic on cities. The COVID-19 pandemic and the associated economic and social impacts have been felt around the world. In large cities and other urban areas, the pandemic has highlighted a number of issues from pressures on urban labour and housing markets, shifts in demographic processes including migration and mobility, changes in urban travel patterns and pressures on contemporary planning and governance processes. Despite Australia's relatively mild COVID exposure, Australian cities and large urban areas have not been immune to these issues. The economic shutdown of the country in the early stages of the pandemic, the sporadic border closures between states, the effective closure of international borders and the imposition of widespread public health orders that have required significant behavioural change across the population have all changed our cities in some and the way we live and work in them in some way. Some of the challenges have reflected long-standing problems including entrenched inequality in labour markets and housing markets; others such as the impact on commuting patterns and patterns of migration have emerged largely during the pandemic. This book, co-authored by experts in their field, outlines some of the major issues facing Australian cities and urban areas as a result of the pandemic and sets a course for the future of the cities we live in.

Brisbane, Australia — Scott Baum
Adelaide, Australia — Emma Baker
Perth, Australia — Amanda Davies
Melbourne, Australia — John Stone
Melbourne, Australia — Elizabeth Taylor

Contents

About the Authors

Scott Baum is trained as an economist and sociologist with research interests in labour market economics, urban sociology and urban and regional development and change. He is currently a member of the Cities Research Institute and the School of Environment and Built environment at Griffith University, Queensland, Australia. He has published widely in the international literature, and his latest book 'Methods in Urban Analysis' was published by Springer in 2021.

Emma Baker is a Professor of Housing Research at the University of Adelaide. Her work examines the health and human impacts of housing and location in urban and regional environments, producing academic, as well as policy-relevant research. Driving this work is a conceptual focus on housing and location as a social determinant of health and an interest in analyses that utilise longitudinal, spatial and administrative big data.

Amanda Davies is a human and population geographer with research expertise in demographics, migration and regional development. She is currently the Head of Social Sciences at the University of Western Australia and has extensive applied research networks with government and industry partners. Amanda's work has been published in leading international journals and cited in government reports and resulting policy related to regional development.

John Stone is a Lecturer in Transport Planning in the Urban Planning Program in the Faculty of Architecture, Building and Planning, University of Melbourne. His research seeks to improve public transport performance in Australian cities through a greater understanding of the professional practice and the political and institutional context for public transport planning in similar cities in Canada and in German-speaking Europe. Since completing his Ph.D. in 2008, his work here and at the Swinburne Institute for Social Research has been funded through grants including an ARC Discovery Project and a Commonwealth Endeavour Research Fellowship. This work follows over 20 years of engagement with public transport management in Melbourne that has included work in local government and the community sector.

Elizabeth Taylor is a Senior Lecturer in Urban Planning and Design at Monash University. Taylor's research explores links between urban planning, housing markets and locational conflict. Often using spatial and historical perspectives, her research develops an understanding of long-term urban change and the role of policy settings in it. She was previously a Vice Chancellor's postdoctoral research fellow in the Centre for Urban Research at RMIT University and a McKenzie fellow at the University of Melbourne, as well as working in consultancy and government research roles.

Chapter 1
COVID-19 and Australian Cities: When the Pandemic Came to Town

1.1 Introduction

As we sit to write these opening lines, it is incredible to think back over the last two years and realise that you never really know what is coming around the corner. January 2020 and Australia had just come off one of the most prolonged droughts in history and was coming to the tail-end of catastrophic bushfires that impacted urban and country people alike. On the political front, our then Prime Minister had gone AWOL to Hawaii for a family holiday. Then in a relative blink of an eye, things changed. And changed in a big way. Although many, including journalist Laurie Garrett (1994), predicted the arrival of a new global pandemic as far back as the 1990s, no one really could have known when, how and what form such a new pandemic might take. Like another addition to the four horsemen of the apocalypse, Australia and the world was now faced with a global pandemic. We didn't know it at the time, but we would be living with the virus for a considerable period.

In late December 2019, the Chinese government notified the World Health Organisation of a mysterious cluster of pneumonia cases in the city of Wuhan in the Hubei province. Cases of the virus, which became known as COVID-19, began to increase—seemingly slow at first, then very rapidly from late January. The Chinese Government responded on 23 January 2020 by rolling out a massive program of testing, contact tracing and home-quarantining of people. Residents of the Hubei province were required to follow strict spatial isolation. Travel, employment, education, recreation and social gatherings were restricted to prevent the spread of the new virus. To some extent, the measures introduced in Hubei province were a sign of what was to come in cities worldwide. While these measures undoubtedly curbed the geographic spread of the virus, by 30 January, following 7,711 reported COVID-19 cases in China and 170 deaths, the World Health Organisation declared the coronavirus a Global Public Health Emergency. It was not until March 11 2020, that they announced the novel coronavirus (COVID-19) outbreak as an international pandemic (Cucinotta and Vanelli, 2020).

S. Baum et al., *Pandemic Cities*, Cities Research Series,
https://doi.org/10.1007/978-981-19-5884-7_1

1.2 The Pandemic Comes to Town

Australia's first case was confirmed when a visitor who had travelled from China to Melbourne on January 19 2020, tested positive. Slowly the number of COVID-19 cases in Australia grew. On the same day as the Melbourne traveller tested positive, it was revealed that three men in New South Wales had also contracted the virus. One was a traveller who had been in Wuhan, and the other two were close contacts. In what seems like a minuscule number now, Australia recorded its 100th case on Tuesday, 10th of March 2020. For context, just over two years on, the country has just passed 6.6 million cases (World Health Organization, 2022). Australians also soon learned that the virus was more than just a simple respiratory illness. Those who caught the disease reported an extensive array of symptoms; headaches, runny nose, fatigue, flu-like symptoms, chest pain and changes in the sense of smell or taste or anxiety. And some people started to suffer from 'long COVID' as their symptoms continued for weeks after being diagnosed. Australia also began to see people dying of COVID. Australia's first death was a 78-year-old man who passed away in a Western Australian hospital on March 1st 2020. He had been a passenger on the Diamond Princess cruise ship, which had been forced to anchor off Japan when the virus was passed among both crew and passengers. As more of the virus emerged, Australia began recording more deaths. Although numerically minor in hindsight, the country recorded its 100th death, an elderly woman in a nursing home, in early May 2020.

As we saw more and more cases emerge, the Australian Government responded by introducing a series of public health measures to help 'stop the spread' and 'flatten the curve'. Daily press conferences were held by state, territory and federal leaders and the heads of the respective health departments to brief communities about the risk, how to deal with the disease and the all-important case numbers. We all learnt a new vocabulary of COVID and pandemics. We learned that 'sanny' was short for hand sanitiser, and people didn't catch the Coronavirus; they got the Rona. Social distancing became the new normal, and face masks and other personal protective equipment became a fashion accessory. Slowly, wearing masks both indoors and outdoors became mandatory, much to the annoyance of those who thought the government was taking away their rights. It seemed like there was free-flow hand sanitiser (sanny) everywhere you went, and the sale of small containers of sanitiser went through the roof as people sought to secure their own personal supply. One of the centrepieces of the public health measures focused on mobility—restrictions to movements across international, state and, in some cases, local government borders caused shock. How could Australian residents caught overseas not be free to return to their country? How could it be that you needed to apply for permission to cross from one state to another?

In a blog post titled "Australia's COVID-19 response: The story so far", Stephen Duckett and Anika Stobart from the Grattan Institute provided what, as we now know, was an optimistic timeline for Australia's response. Starting with small piece-meal actions such as restricting travel to and from certain countries and requiring overseas arrivals to self-isolate, approaches soon ramped up as case numbers increased exponentially. Sadly, despite the suggestion of a "transition to the new normal" in late April 2020, we have, as we know, continued to face challenges as new variants of the disease have emerged (Fig. 1.1).

The measures enacted during the escalated national action and subsequent periods by state, territory and federal governments to flatten the curve were actions to address the immediate physical health threat. Still, at the same time, these actions acted to interrupt urban life significantly. Tens of thousands of Australians became out of work, and the bulk of the county's 24.6 million people were told to stay home unless they absolutely had to go outside. The actions changed the way people interacted with others and the places people frequented, even in their local community. Photos started appearing on social media portraying eerily deserted streets. In one newspaper post, Sydney's usual busy CBD was shown to be practically empty (Fig. 1.2). Other cities across the country were the backdrop to similar visions. Several local governments even went to the extremes of taping off playground equipment because of the potential for the virus to spread via contact with 'high-touch' items like swings and slides (Fig. 1.3). The pandemic had impacted all aspects of urban life: health access, employment, public space, education and social inequality.

Like almost all countries, Australia's large cities bore the brunt of the rise in cases. For many non-metropolitan regions, the first cases of COVID-19 didn't arrive

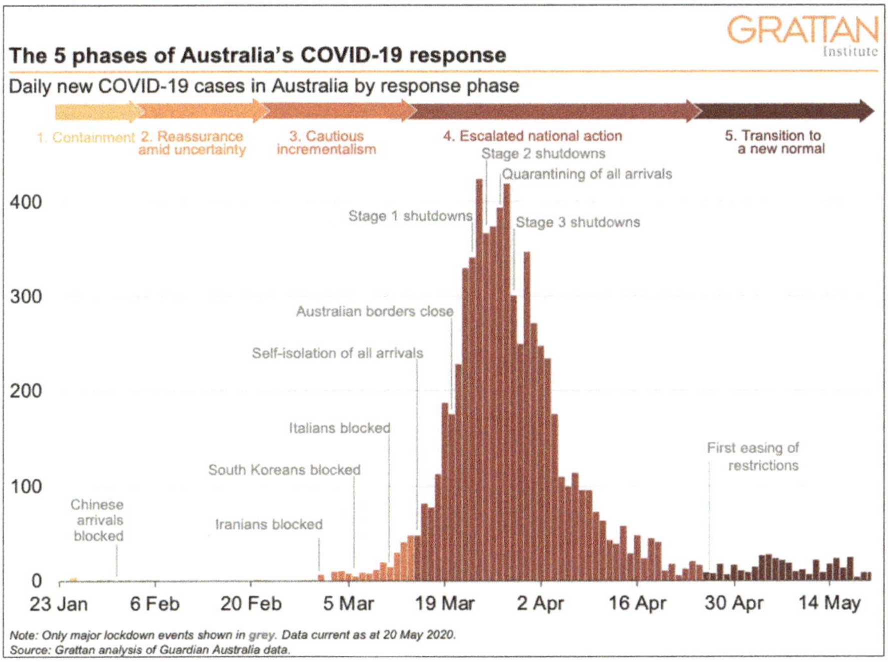

Fig. 1.1 The 5 phases of Australia's COVID-19 response (Duckett and Stobart, 2020)

Fig. 1.2 A deserted Sydney CBD (Herde, 2022)

Fig. 1.3 Playground closure (De Poloni, 2020)

until the spread of the omicron variant in 2021/22. During the early stages, questions were naturally being asked about why was COVID-19 hitting some places harder than others? What were the characteristics of places that made them more vulnerable? Was it something to do with the population characteristics or something else? What facts shape the spread of the disease across different locations in the same city? Some of the answers were straightforward. Large cities had significant international linkages meaning that these global nodes were the backdrop to initial and subsequent influxes of disease. Other responses were more nuanced, and our understanding developed as the pandemic unfolded.

The questions swirling around were important, and policy responses were often ad-hoc and developed seemingly on the run. The lockdown of nine public housing towers in Melbourne over fears of the 'explosive potential' to spread the disease was based on the understanding that population density was a key cause of the spread. At the time, officials were quoted as saying.

> We know one of the biggest weapons in stopping the spread of coronavirus is keeping our distance, and that the virus has more chance of transmission indoors.
>
> Tenants in big apartment blocks often share facilities like lifts, corridors, rubbish facilities and laundry rooms—making the chance of running into someone higher.
>
> Each one of the Melbourne tower blocks is home to hundreds of residents living in small units without balconies.
>
> Those towers have a large concentration of people in a small area,
>
> And so we have to take particular notice and particular attention to make sure that the spread is minimised and that people are protected (Murray-Atfield, 2020).

It also became apparent that the disease was not blind to the socio-economic cleavages that are an endemic characteristic of cities and society. The fact that the disease was spread by close contact meant that race, culture and socio-economic status became an adjunct to understanding its spread as different groups of people and communities were identified as being at higher risk due to their employment circumstances or domestic living arrangements. These understandings appeared to come later and often to the detriment of those most impacted. The lockdown of the Melbourne apartment towers was seen by many as reflecting race, culture and socio-economic status biases. Media reports began appearing that vilified certain groups or individuals. In one such case, an Adelaide Medi-hotel security guard who had lied to contract tracers about his movements and his second job at a pizza shop was called out by many, including the South Australian Premier:

> To say I am fuming about the actions of this individual is an absolute understatement. The selfish actions of this individual have put our whole state in a very difficult situation. His actions have affected businesses, individuals, family groups and is completely and utterly unacceptable (Kurmelovs, 2020).

While the premier and others were quick to blame this individual, others pointed to the fact that the case was more a reflection of the government's failed COVID responses, the impact of widely unequal employment opportunities for many and the need for low-income workers to hold more than one job. As economics blogger Bill Mitchell wrote, "Why improve policy when a government can pillory a low-paid, precarious worker instead" (Mitchell, 2020). Others pointed out that had the person not been a migrant but rather a white middle-class Australian, the treatment and fall-out would have been somewhat different.

Understanding how the virus was spread and its impacts have become an industry all of its own. Estimates from mid-2020 suggested that scientists had published 7,500 articles on COVID-19, and economists had produced over 110 working papers and the odd book (Nathan, 2020). Apart from the science and epidemiology of the disease, there was a growing interest in exploring the social and economic aspects of the pandemic response. How did governments' policy responses play out on the ground? What did forced lockdown do to the vibrancy of cities and other large urban areas? What would be the long-term impacts on cities around the world?

1.3 About This Book

These types of questions are where this book picks up the COVID-19 baton. In the chapters that follow, we address some of what we see are the significant issues impacting Australian cities in the wake of the pandemic. How did the pandemic impact the populations within cities, were some communities affected more than others? What about work? How did the arrival of COVID-19 impact the labour markets across Australian cities? How did housing play a part in the pandemic? What about transport or how we plan our cities? Even though we were finishing up the chapters for this book, the pandemic was still with us. We wrote chapters during various stages of lockdown and with some of us suffering through the dreaded Rona as we watched the unfolding impacts on our urban areas and communities.

In considering the impact of the COVID-19 pandemic on Australian cities, we have all focused on our individual and collective academic interests. As such, our personal biases have influenced what is included and how we have interpreted the patterns and outcomes. We have focused our attention on what we considered to be the main vital issues—inequality, transport, housing, employment, demography and planning. For us, these are the main bits of urban life that the pandemic has impacted. They might not be the only ones, and others will likely disagree with our focus. But they will at least get people thinking about what has been happening in our large population centres over the past couple of years and perhaps begin to think about what we might do better going forward.

There will also be some overlap between what we have identified and discussed in Australian cities and the outcomes in other cities worldwide. There will also be marked and significant differences. Socially and economically, city structures differ, levels of development and infrastructure differ and city planning and politics differ. What we present here will hopefully be fed into a broader global discussion that leads to better understanding.

The COVID-19 pandemic has thrown up a multitude of challenges for cities and the people that call them home. These challenges have been met with varying levels of success as governments, policymakers and citizens alike have grappled with the unknown. Hopefully, like all challenges, we learn something from the past couple of years.

References

Cucinotta D, Vanelli M (2020) WHO declares COVID-19 a pandemic. Acta Bio Med : Atenei Parm 91(1):157

De Poloni G (2020) How coronavirus could forever change the look and feel of Australia's cities and suburbs. *ABC News*. Retrieved from https://www.abc.net.au/news/2020-04-11/how-coronavirus-could-forever-change-our-cities-and-suburbs/12137122

Duckett S, Stobart A (2020) Australia's COVID-19 response: the story so far Retrieved from https://grattan.edu.au/news/australias-covid-19-response-the-story-so-far/

Garrett L (1994) The coming plague: newly emerging diseases in a world out of balance: Farrar, Straus and Giroux

Herde C (2022) Google data shows shoppers and workers avoiding Australian CBDs as Omicron sweeps the country. *The Australian*. Retrieved from https://www.theaustralian.com.au/business/google-data-shows-shoppers-and-workers-avoiding-australian-cbds-as-omicron-sweeps-the-country/news-story/8f305c8bb52132b1d273c204045759ab

Kurmelovs R (2020, 23 November). South Australia makes pizza worker scapegoat for Covid-19 failures. *The Guardian*. Retrieved from https://www.theguardian.com/world/2020/nov/21/south-australia-makes-young-pizza-worker-scapegoat-for-covid-19-failures

Mitchell W (2020) Why improve policy when a government can pillory a low-paid, precarious worker instead. Retrieved from http://bilbo.economicoutlook.net/blog/?p=46402

Murray-Atfield Y (2020) Why Melbourne's public housing towers have 'explosive potential' for coronavirus to spread. Retrieved from https://www.abc.net.au/news/2020-07-06/why-melbourne-locked-down-public-towers-are-a-coronavirus-worry/12423934

Nathan M (2020) The city and the virus. *Urban Studies*, 00420980211058383

World Health Organization (2022) WHO coronavirus disease (COVID-19) dashboard. Retrieved from https://covid19.who.int/region/wpro/country/au

Chapter 2
Population Challenges and Change in Pandemic Cities

2.1 The Big Population Questions for COVID Cities

Our opening discussion clearly shows that COVID-19 has had wide-ranging impacts across Australian cities' social and economic landscape. But the arrival of the pandemic and the government response to it also had fundamental implications for the population dynamics underway within cities. As the disease spread, population growth was disrupted while demographic structures and the spatial distribution of the global population changed. For many countries, COVID has resulted in an increased mortality rate, with questions now being asked about the long-term impacts on the demographic structure, fertility rate and ageing (Harper 2021).

It is true to say that Australia has largely avoided the high COVID-19 mortality rates witnessed elsewhere. However, this does not mean that the country hasn't faced its share of population challenges as public health measures were introduced to curb the spread of the disease. While Australia's public health response to COVID-19 has undoubtedly resulted in Australia being in the enviable position of not having to substantively revise lifetables due to increased mortality, disruptions to international and internal migration have driven some shifts in Australia's population demography and geography.

With more than two-thirds of Australia's population concentrated in just seven capital cities and about 90 per cent of the people living in an urban setting, it should be no surprise that the population impacts of the pandemic have shaped the lives and livelihoods of those living in cities. As the implications of the pandemic are being realised, the most pressing population questions to emerge for cities were:

- How did COVID-19 and related public health measures impact short-term and long-term international migration flows to and from Australian cities, and
- How did COVID-19 and related public health measures impact internal migration from to and from cities?

Bundled within these two broad areas of concern, the pandemic has given rise to numerous hypotheses about the changing nature of mobility within Australia.

S. Baum et al., *Pandemic Cities*, Cities Research Series,
https://doi.org/10.1007/978-981-19-5884-7_2

Initially, in response to the unprecedented closing of the international border, attention was given to what impact returning ex-pats might have on the job and housing market—would the job market be saturated, and would the housing affordability crisis intensify? As the pandemic became entrenched in our large cities, and many started having to work fully or partially from home, questions were raised about the future of central workplaces and, in turn, the utility of central business districts. Somewhat following this, as social commentary increased about how working from home arrangements had given workers improved capacity to work anywhere, people began asking if people would move away from 'unsafe' cities in favour of 'safe' rural locations and lifestyles. We started seeing newspapers reporting 'testimonies' of individuals who had relocated from capital cities to regional areas and, in turn, 'testimonies' from regional business and estate agents reporting on the impacts of the so-called 'city exodus' (Terzon 2021). An article in The Age (2021) summarised these thoughts.

> Prompted by the pandemic, tens of thousands of Melburnians have upped sticks in the past two years and moved to regional Victoria in search of more affordable housing, clean air, community and a better quality of life for their children. For some, the trigger would have been the six successive lockdowns, being cooped up in the city or the suburbs with nowhere to go and nothing to do — all the negatives of city life with none of the benefits (para 1).

When writing these pages, the Australian population is still experiencing the effects of the pandemic, with considerable uncertainty remaining about the longer-term implications of the pandemic. However, delving into the demographic and population data that emerged during the first two years of the pandemic and considering how these might have differed from the trends if the pandemic hadn't arrived on our shores provides insights into the short-term changes and likely longer-term trends.

2.2 Fortress Australia and International Migration

The most apparent immediate impact of the COVID-19 pandemic on Australia's population was the reduction in international net migration. Early in the pandemic, the federal government decided to introduce new international border control restrictions, significantly restricting the movement of visitors to Australia and Australian's movements to and from the county. The international border restrictions were quickly termed a 'hard border' as, for most, the country effectively became closed to all international travel. In early March 2020, if you were an Australian overseas, the Department of Foreign Affairs and Trade (DFAT) told you to return home ASAP ahead of possible COVID-19 border closures (Martin and Murphy 2020). The strict closures arrived mid-way through March 2020. Monthly arrivals for each state and territory were capped, and overseas travel from Australia was (mostly) banned (with some commentators suggesting that the exception to the international travel ban was for those categorised as very wealthy or famous).

For Australia, this temporary reduction of international migration had the potential to be very significant, not least because of its impact on our level of population growth. Before the pandemic, the Australian Bureau of Statistics predicted that over the coming decade, 0.7–1.1% of Australia's per annum population growth was projected to be the result of net international migration (Australian Bureau of Statistics 2021c). Running the numbers, if total population growth for Australia over the next decade was projected to reach somewhere between 0.9 and 1.4% per year, international migration was projected to account for more than 60% of Australia's projected growth (Australian Bureau of Statistics 2021c). This potential decrease in the number of people coming to call Australia home had implications beyond simply messing with the Bureau of Statistics' projections. Lower population growth relates to less demand for things (housing, schools), reduces labour supply and, as migrants are typically younger, fails to offset the ageing population (Centre for Population 2021).

Much of what we have seen regarding the rate of international migration to Australia and the spatial distribution is substantively the result of government policy limiting the number of arrivals for different visa categories. In 2018/19, before the pandemic, Australia's net international migration was around 240,000 people, of which 40% were visitors and tourists coming to see our sights, 26% were migrants planning to settle in Australia, and 14% were young people (generally) coming to study in one of our education institutions (Australian Bureau of Statistics 2021b). Once the international borders closed, net international migration during 2019/20 fell to 194,400 (Australian Bureau of Statistics 2021b). That period included three months where the COVID-19 response border restrictions were in place. For 2020, net international migration totalled only 3,250, reflecting the crash in international travel during the second half of 2020 when COVID-19 restrictions were beginning to bite (Australian Bureau of Statistics 2021a).

International migration is vital to Australia's financial bottom line. Recognising the impact of the hard border, in preparing the budget for 2021/22, forecasters estimated international migration for 2020/21 to be about minus 72,000 (Love and Spinks 2021)—a massive hit on previous forecasts and budget plans. Signalling a clear recognition of the importance of international migration to Australia's financial wellbeing, the 2021/22 budget was developed on the assumption that a small level of international travel to Australia, particularly for international students, would commence in late 2021. It was also assumed that broader international travel and permanent migration to Australia would resume from mid-2022. Based on these assumptions, net overseas migration to Australia was forecast as −21,600 for 2021/22, increasing to 95,900 during 2022/23 and further increasing to 201,100 during 2023/24 (Love and Spinks 2021). International travel to Australia resumed on 21 February 2022, a little later than anticipated in the 2021/2022 budget. However, the impact of the opening up of fortress Australia is unclear. At the time of writing, only fully vaccinated people can travel to Australia, and the normal visa restrictions still apply.

So given what we know, how did the decline in net international migration impact Australia's cities? What was the geography of the limited population flow into Australia? Firstly, it is important to realise that population growth from net international migration is not evenly distributed across Australia. Over the last decade,

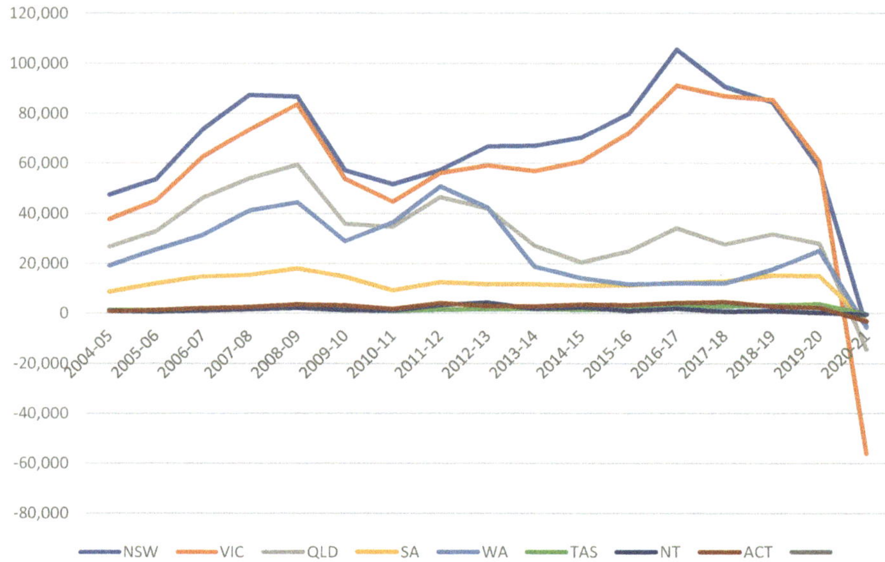

Fig. 2.1 International migration by state, 2005–2021 (Australian Bureau of Statistics 2021a)

the two largest states of New South Wales and Victoria received the substantive proportion of population growth from international migration (Fig. 2.1). But all of that disappeared as COVID-19 border restrictions hit.

The potential impact of international migration flows on Australian cities become more apparent when we consider the urban and regional imbalance within each state. We know that new international migrants prefer capital city locations over regional locations. In Victoria, 92% of international migrants live in Greater Melbourne, while in New South Wales, 86% live in Greater Sydney (Australian Bureau of Statistics 2021c). This pattern is also true for Australia's other states and territories, except for Queensland (Australian Bureau of Statistics 2021c). In Queensland, 40% of international migrants locate outside of Greater Brisbane. Importantly, though, the large metropolis of the Gold Coast, adjacent to Greater Brisbane, attracts a large share of this international migrant population (Australian Bureau of Statistics 2021c).

During the pandemic, what has been unclear is if the loss of international migration will be evenly distributed across this already spatially uneven pattern of distribution—or will the loss be more proportionally concentrated in regional or city areas or some states? It is difficult to say, but if we compare actual population growth numbers with those same numbers less the net overseas migration, we can get some idea. There is no doubt that the impact was significant. Capital cities, whose population growth is so dependent on net overseas migration, witnessed a difference of 1.2 percentage points once the migration figures were removed (Fig. 2.2). In absolute numbers, this equates to a little over 200,000 fewer people (Australian Bureau of Statistics 2020)! The impacts are seen across all cities, but as suggested by the Treasury Department's Centre for Population (2021), the fall in net overseas migration would likely hit

Greater Sydney and Adelaide the most, resulting in meagre population growth in these cities (Fig. 2.3).

The Centre for Population data also provides some insights into how changes in net overseas migration might impact Australian cities' internal dynamics and structure. Again, population growth plummets with the absence of overseas, with the most significant impacts being in the inner and middle-ring suburbs of the large capital cities (Fig. 2.4).

The bulk of net international migration comprises those people holding some form of temporary visa, with students and holidaymakers being granted the most visas (Australian Bureau of Statistics 2021b). It is no wonder that when the Australian government announced the closure of international borders and limited who could

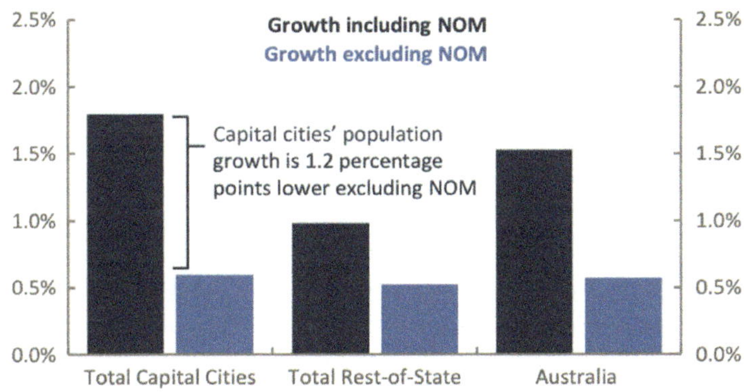

Fig. 2.2 Population growth with and without NOM, 2018–19 (Centre for Population 2021)

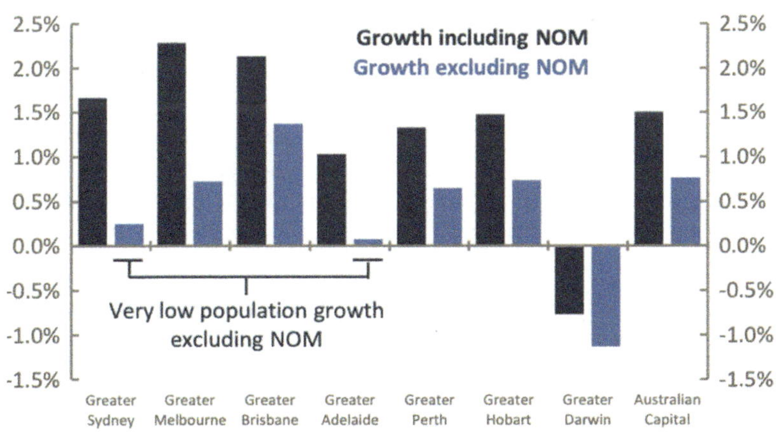

Fig. 2.3 Population growth by capital city with and without NOM, 2018–19 (Centre for Population 2021)

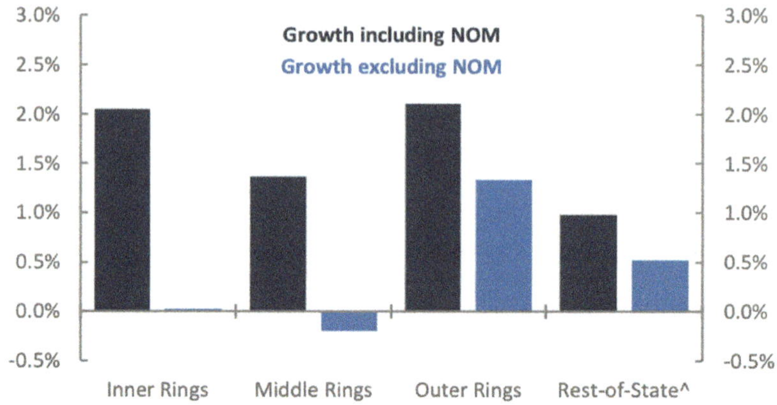

Fig. 2.4 Population growth of the five largest capital cities by city rings, and all rest-of-state areas, with and without NOM, 2018–19 (Centre for Population 2021)

enter the country, many commentators were quick to point out the potential negative impacts this would have across a broad range of sectors. Most notably, and unsurprisingly, we witnessed almost immediate economic impacts on tourism and higher education sectors and significant effects on the vibrancy of particular city spaces. Student precincts in cities became less lively, and the tourist zones around the country became decidedly deserted.

But just how much did numbers for these different migration flows drop? Take the sudden fall in international students. Government figures show that for the April to June financial quarter, international student visa applications had fallen by 61% from the previous year. We know that the majority of international students gravitate towards Victoria (40%) and New South Wales (31%), with much smaller numbers in Queensland (11%), Western Australia (8%) and South Australia (6%). Given this underlying unevenness, the ongoing disruption to international student migration flow would disproportionally impact Victoria and New South Wales. Indeed, we can see the immediate impact of COVID-19 on temporary student migration to Australia in 2020. Interestingly, while both Sydney and Melbourne attract the largest share of international students (Hurley 2020) due to the concentration of higher education providers and facilities in Melbourne's CBD, this loss for businesses and accommodation providers was most significantly felt in the inner city. The impact was more spatially distributed across Greater Sydney, reflecting the more disbursed geography of tertiary providers and accommodation.

Skilled migration and family migration flows are also geographically variable. Skilled migrants necessarily move to localities where their employment is located, and, depending on the visa category, some move to regional areas. Family migration builds on recent migration flows, creating distinct spatialities for diasporas. While the border closure resulted in a sharp reduction in migration in both of these categories, it remains unclear what the longer-term implications might be. This is because the drivers of international migration are impacted by both events occurring in Australia and outside of Australia. For family migration, given that the factors underpinning this are largely associated with the individual and their family networks, it is plausible that this category would recover relatively quickly. Indeed, with restricted global travel, it is possible that there could be increased demand for family visas. However, it is unknown how regulations around vaccination, for example, might impact the ability of potential in-migrants to move to Australia.

While it is clear that interruptions to international migration have had short-term impacts on Australia's population and that this impact has been spatially variable, it is too early to determine with any reliability what the trends will be long term. As noted above, Australia's 'bounce back' is not simply determined by factors in Australia. The social, economic and political circumstances in sending countries will also impact migration flows. If Australia is seeking to recover its international migration flows, targeted strategies will need to be developed that address any place-based barriers to migration that might arise as well as any COVID-19 related governance arrangements that emerge.

2.3 Sea Change, Tree Change, or No Change: Internal Migration

It is not just people crossing our international borders that are important to understand the COVID population questions. People move around the country too. And often for a whole range of reasons. For many years now, there have been talks about people moving from Australia's cities to regional towns—typically categorised as a sea change or a tree change—in order to pursue more fulfilling and healthy lifestyles. The arrival of COVID-19 bought its own take on the question of urban to regional migration. Stories about population loss from cities and potential shifts in Australia's internal migration have generated considerable public and media interest. Indeed, particularly eye-catching was the August 2021 update from the Australian Bureau of Statistics (2021d) that reported, "Capital cities had a net loss of 11,800 people from internal migration, *the largest quarterly net loss on record*" (para 4, emphasis added). This statement followed one made by the ABS in February 2021 about international migration in the September 2020 quarter, which read, "Capital cities had a net loss of 11,200 people from internal migration, the largest quarterly net loss on record" (para 8). Quickly reported by the media, headlines of city population decline further underpinned the thesis that COVID-19 has resulted in Australians seeking to leave city life for the rural idyll—made all the more possible by the rapid transition within workplaces to online working arrangements (see for comment Davies 2021).

It is clear that since the commencement of Australia's public health responses to COVID-19, internal mobility has been curbed. As state governments declared COVID hotspots, state and regional hard borders were put in place, effectively limiting or curtailing interstate and, during some periods, intrastate migration. Western Australia famously became known as Australia's hermit state, with unkind comparisons with North Korea. Stories of people stuck at state borders with moving vans full of furniture became legendary. But where public health rules allowed, people were still moving between and around states.

It is true that since the COVID pandemic, some capital cities have lost population through internal migration, and some regional areas have experienced population growth. However, this should not be interpreted as rural population growth or a shift in Australia's dominant urban-centric settlement pattern. Indeed, some capital cities have experienced population growth. To understand what is going on and what longer-term trends might have emerged, it is essential to look beyond the headline 'net' data and, in particular, consider shifts in arrival and departure data.

Our largest metropolitan region, Greater Sydney, recorded the greatest loss of population through the internal migration of any capital city since the onset of COVID-19. But, this is not necessarily a COVID thing. Sydney had recorded a net loss each quarter since September 2001—which is the first recorded period of the internal quarterly migration data series produced by the Australian Bureau of Statistics. Furthermore, the actual quarterly loss from Greater Sydney has long been variable—moving from a low of approximately 4,000 to a high of roughly 11,000. In 2002, 2003, 2004 and 2018, the quarterly loss exceeded the total quarterly loss recorded during COVID-19. Finally, there is no consistent pattern for the variability in the net quarterly loss. Some periods are driven by decreased in-migration from interstate or intrastate, while others are driven by increased outmigration.

While during the pandemic, there has been little change from the long-term pattern of net outmigration, what has occurred is that there has been an increase of approximately 2% in the number of people leaving Greater Sydney for intrastate locations and a 3% decline in the number of people moving to interstate capital cities. Of those who left Sydney during the pandemic, approximately 45% moved to regional NSW, with a further 21% moving to the capital cities of Brisbane (when the Queensland Government did not shut them out) and Melbourne. Despite these small shifts, it is not currently the case that proportionally more people are now moving from Greater Sydney to regional NSW than previously, with higher rates of regional migration recorded in the early 2000s.

Considering the age of those who departed Sydney during the pandemic, there has been no change in the overall age profile of the departing population. The largest group leaving were those aged 25–44 years, with the second-largest group being those aged 45–64 years. Of the departing population, only between 7 and 10% were aged 65 and over, which is an underrepresentation of the older population. Considering the age profile of those who moved into regional NSW during the COVID-19 period, there was also no change—suggesting that it is unlikely that, at this point, there has been a significant change in the drivers underpinning regional migration patterns across NSW as a result of COVID-19.

Our second-largest metropolitan region, Greater Melbourne, recorded the second largest overall net loss of population through internal migration during the pandemic to date. Unlike Sydney, the net loss of people from Greater Melbourne was unprecedented. Since 2012, Greater Melbourne has enjoyed net population growth through internal migration. However, in 2018, Melbourne slipped into a net population loss phase with a total loss of just under 1,900 people in 2018 and just over 500 people in 2019 due to internal migration. This decline was driven by a net reduction in the number of people moving to Melbourne from interstate destinations and a steep increase in the number of Melbourne residents moving interstate. Mobility between Greater Victoria and Melbourne has remained steady as a proportion of total movements.

As for Sydney and NSW, there was little change in where people were moving to and from during the pandemic regarding both Greater Melbourne and regional Victoria. Of people leaving Greater Melbourne, approximately 40% moved to regional Victoria, with a further 20% moving to the capital cities of Brisbane and Sydney. When considering the age of those who departed Melbourne during the COVID-19 period, again similar to Sydney, there has been no noticeable shift in established trends when considering age structure. However, reflecting on the observation that there has been a net increase in people leaving Melbourne during the COVID-19 period, it is the case that this increase occurred across all age groups—older people were not more or less likely to move out of Melbourne than working-age people.

Unlike Melbourne and Sydney, Greater Brisbane has been experiencing population growth during the pandemic. However, this growth trend follows 21 quarters of net growth, driven by strong interstate migration. During the pandemic, this established trend of interstate migration continued, with the major sending locations outside of Queensland being Melbourne, Sydney and regional New South Wales. While there was little change in the dominance of sending locations, the number of people moving from Melbourne to Brisbane did double as a proportion of total movers. The number of people moving to regional Queensland during the period also increased. Indeed, during the pandemic, the number of people moving into regional Queensland from interstate locations was the highest since 2004. Following longer-term migration trends to Queensland, it is expected that much of this in-migrant population moved to the large urban areas of the Gold Coast and Sunshine Coast.

With Melbourne experiencing extended periods of 'lockdown' in 2020 and 2021, it can be speculated that these periods of extended lockdown in Melbourne may have contributed to the observed increase in the number of residents leaving Melbourne and Victoria. While there was an increase in the number of people moving, no clear trends have emerged to suggest that there has been a change in the established geography of migration flows. However, this noted, it is recognised that there has been an overall proportional increase in Melbourne residents moving to Brisbane and regional Queensland since the start of the pandemic. At this point, it is difficult to determine how much of this growth can be attributed to a natural increase in domestic migration driven by social and economic factors in Queensland (such as a strong jobs market and relatively lower costs of housing) or attributed to COVID-19 related 'push factors' in Melbourne.

Like Brisbane, Perth has experienced net population growth through internal migration, moving the focus to the west coast. For Perth, there has been a clear shift in internal migration since COVID-19, with the city experiencing growth predominantly due to a decline in interstate departures. Regional Western Australia has also experienced a decrease in people leaving for interstate. Regional Western Australia is moving into a population gaining phase, which is typically only observed during periods of solid job growth in the State. Of those who did leave Perth and Western Australia, the majority moved to Melbourne and Sydney. It is probable that border closures and periods of lockdown in Melbourne, in particular, together with the relative strength of the job market in Western Australia, contributed to the shift in internal migration trends for Perth and regional Western Australia.

For South Australia, there was a slight decline in population loss due to decreased outmigration from Adelaide. Similar to Perth, this shift probably reflects the relative strength of the job market in Adelaide and, with Melbourne a key receiving destination, perhaps the social and economic disruptions in Melbourne related to extended lockdowns.

Internal migration patterns for the Northern Territory, Australian Capital Territory and Tasmania have not remained static during the COVID-19 period. However, the volatility from quarter to quarter in internal migration reflects longer-term trends in these jurisdictions, and it is not necessarily possible to determine how or if COVID-19 has impacted internal migration trends.

Overall, Australia's internal migration data suggests a shift in some internal migration trends since public health measures were introduced to curb the spread of COVID-19. Most notability, Melbourne, which endured the most extended period of lockdown of any jurisdiction in Australia in 2000, and the again experienced extended lockdown in 2021, had a greater than average rate of outmigration and, conversely, greater Brisbane and regional QLD, which largely avoided periods of extended lockdown, experienced increased in-migration from Melbourne. Like Brisbane, Perth remained relatively COVID-19-free during 2020 and 2021 and underwent a population turnaround, with the State moving into a population gaining phase, largely the result of decreased outmigration. This suggests that the pull factors that typically drive net interstate population loss from Perth and Western Australia have declined. As noted above, the relative strength of both the Queensland and Western Australian economies and job markets is also likely related to population growth in both states. Tellingly, the ABS's Household Impacts of COVID-19 Survey for June 2020 found that for 94% of Australians, COVID-19 did not influence their plans to move over the next 12 months, 4% were less likely to move and only 2% were more likely to move (ABS, 2020).

2.4 Where to Next for Australia's Population

For life in Australian cities, international migration has been critical to shaping the economic, cultural and built landscapes. The pandemic has had clear and immediate impacts on Australia's population growth, with the decline in international migration slowing the population growth rate and increasing the structural ageing rate. The effect has been spatially uneven and, in particular, the temporary loss of students and international visitors impacting the inner-city areas of Melbourne and Sydney. What is yet unknown is how fast international migration will recover and, more specifically, how fast different categories of international migration will recover. This is because international migration results from factors occurring in both sending and receiving areas—and COVID-19 has changed these. For example, considering international students, with the UK and Canada opening their borders almost a year before Australia, these destinations will have attracted a portion of the flow of international students that normally would have come to Australia's cities.

Internal migration flows also shape life in Australian cities. Changes in internal migration patterns attracted a lot of media attention, particularly the increase in net outmigration from major capital cities, with some questions being raised about the safety and desirability of major cities. Since the pandemic's start, there have been some changes to internal migration that could be attributed to factors related to the pandemic, such as border restrictions and the relative strength of the jobs market and economy in various states. There is the potential that the pandemic has resulted in some people moving forward with their decisions to move out of capital cities for employment, retirement or other reasons. For others, it is plausible that the pandemic and the public health response, particularly border closures and lockdowns in major cities, have delayed their decision to relocate to Sydney or Melbourne for study, employment, or other reasons. However, given Australia's established trends of internal mobility, it is most probable that for those that have undertaken domestic moves during the pandemic, it is the result of a combination of social, economic and personal factors that operate at both the sending and receiving destinations that are not directly related to the pandemic. While working from home arrangements for some segments of the labour force has delivered more flexibility about where to live, for the majority of the population, this flexibility does not mean they can physically relocate permanently to another location. More than just employment ties people to places, including friend and family networks and housing investments. For those with children or others, they care for in the home, the education and social activities of those they care for, as well as access to health resources, are also essential considerations impacting migration decisions. Therefore, while COVID-19 is generating options for some to relocate, there is no evidence to suggest that COVID-19 is underpinning a significant change in where Australians live.

It is too early to draw conclusions about any longer-term changes to international or internal migration trends, their impacts, or the causal relationship to the pandemic. While there are some short-term impacts, and at this point, the impacts are relatively minor, there is a need for population policies and policies that inform population distribution to be agile so as to respond to any shifting migration trends both within Australia and those external to Australia that have implications for population growth and distribution across Australia's cities.

References

Australian Bureau of Statistics (2020) Regional Population, Statistics about the population and components of change (births, deaths, migration) for Australia's capital cities and regions, 2018–19 financial year. Retrieved from https://www.abs.gov.au/statistics/people/population/regional-population/2018-19

Australian Bureau of Statistics (2021a) Net overseas migration, arrivals, departures, state/territory, major groupings and visa—calendar years, 2004 to 2020. Retrieved from https://www.abs.gov.au/statistics/people/population/migration-australia/latest-release#data-download

Australian Bureau of Statistics (2021b) Net overseas migration, arrivals, departures, state/territory, major groupings and visa—financial years, 2004–05 to 2019–20. Retrieved from https://www.abs.gov.au/statistics/people/population/migration-australia/latest-release#data-download

Australian Bureau of Statistics (2021c) Population projections. Retrieved from https://www.abs.gov.au/statistics/people/population/population-projections-australia/2017-base-2066

Australian Bureau of Statistics (2021d) Regional internal migration estimates, provisional: March 2021d. Released 03 Aug 2021d. Retrieved from https://www.abs.gov.au/statistics/people/population/regional-internal-migration-estimates-provisional/latest-release

Centre for Population (2021) Low overseas migration: a quick guide to the potential impacts on local populations. Commonwealth of Australia Retrieved from https://population.gov.au/sites/population.gov.au/files/2021-12/low_overseas_migration_quick_guide.pdf

Davies A (2021) Has COVID really caused and exodus from our cities? In fact, moving to the regions is nothing new. The conversation. Retrieved from https://theconversation.com/has-covid-really-caused-an-exodus-from-our-cities-in-fact-moving-to-the-regions-is-nothing-new-154724

Harper S (2021) The impact of the Covid-19 pandemic on global population ageing. Journal of Population Ageing 14(2):137–142

Hurley P (2020) Coronavirus and international students. Retrieved from https://vuir.vu.edu.au/42245/1/international-student-update-2020-mitchell-institute.pdf

Love S, Spinks H (2021). Immigration: budget review 2020–21 index. Parliament of Australia Retrieved from https://www.aph.gov.au/About_Parliament/Parliamentary_Departments/Parliamentary_Library/pubs/rp/BudgetReview202021/Immigration

Martin S, Murphy K (2020) Government urges Australians overseas to return home as borders close around the world. The Guardian. Retrieved from https://www.theguardian.com/business/2020/mar/17/coalitions-second-multibillion-dollar-coronavirus-stimulus-expected-to-target-businesses-and-low-income-earners

Chapter 3
COVID-19 and the Social Structure of Cities: The Forgotten Vulnerable

3.1 Introduction

> On various occasions, the phrase "COVID-19 does not discriminate" has been repeated. This, however, is a dangerous myth, sidelining the increased vulnerability of those most socially and economically deprived (Patel et al. 2020, p. 110).

This quote from the academic journal Public Health helps us frame the COVID-19 health pandemic within its broader social context. Like all health issues, the pandemic is an emergency that has deeply entrenched social connections and implications. The main point is that the health outcomes associated with the pandemic are not evenly shared between different individuals or communities. All through the pandemic, but especially in the early phases, it has been those with lower incomes, older people, people with poor English skills and those with existing health issues that have been at the mercy of the disease. The association between these population characteristics and the risk of contracting COVID-19 are part of the so-called social determinants of health-the non-medical factors that influence health outcomes and account for the unfair and avoidable differences in health status. They are often considered at the level of individuals but, as we will see, come out to play in the socio-economic spaces of cities. These social determinants are what the Australian Institute of Health and Welfare refers to when they note that some groups in the population were more affected by COVID-19 than others:

- In 2020, 7% of all COVID-19 cases in Australia and 75% of all deaths were in people living in residential aged care facilities.
- Up to early July 2020, it was estimated that Australia's healthcare workers were 2.7 times as likely to contract COVID-19 as the general community.
- There were almost four times as many deaths due to COVID-19 for people living in the lowest socio-economic group compared with the highest socio-economic group. Age-standardised mortality rates were 2.6 times as high (Australian Institute of Health and Welfare 2021, p. vi).

S. Baum et al., *Pandemic Cities*, Cities Research Series,
https://doi.org/10.1007/978-981-19-5884-7_3

Extrapolating from these kinds of figures allows us to start thinking about the processes that have led to the uneven nature of COVID-19 outbreaks. Given the viral nature of the disease, the most important social functions are associated with maintaining social distance and following public health mandates. The most disadvantaged people in our society are overwhelmingly more likely to live in overcrowded dwellings, often without any choice. In Australia, overcrowding is an issue played out in our cities (Herath and Bentley 2018a), with media reports describing situations of "10 people in a two-bed unit", "58 beds crammed into 19 dirty, makeshift rooms", and "10 people shoehorned into one bedroom, tenants sleeping in bathrooms, and in one case, a pantry" (Herath and Bentley 2018b). As we have known from the beginning of the pandemic, being in close proximity to others does not bode well for disease suppression and so living in crowded environments adds to the disease risk.

But it is not only indoor space that puts disadvantaged city dwellers at risk. Often associated with overcrowding, but not always, is the limited access to open outdoor space, either privately or at a community level (think open green space). The outcome is? Those with poor access to outdoor space are less likely to adhere to social distancing measures. Not because they don't want to or are partaking in some sort of civil disobedience, but because it is just more difficult for them.

Disadvantage also haunts people in other ways, upping their exposure risk. Throughout the pandemic, it has been increasingly clear that workers, in particular, fields are more likely to face exposure to the virus. These are front-facing occupations, including nurses, security guards, cleaners, supermarket attendants and the increasing numbers of gig economy food delivery drivers and riders who, because they cannot work remotely (we talk about this in Chap 4), are more likely to be exposed to the virus. The problem, which became evident in the early stages of the outbreak, was that for many of these workers, low levels of remuneration meant that when a worker did get infected, they felt compelled to return to work and not remain in isolation because of the need to continue to support their families or themselves. Again, a perfect storm whereby an individual's socio-economic (in this case, economic) circumstances put them at greater risk of contracting the disease.

Then there is also the question of understanding the health situation as it emerges. One of the most difficult challenges during the pandemic was effectively communicating important information to a broad range of communities with different challenges. We know that approximately 4% of the Australian population have poor English skills. This represents around 800,000 people who are potentially missing crucial information. Added to this is the approximately 14% of Australians who have poor literacy levels and may not be able to read material that requires more than a basic level of education (Australian Academy of the Humanities 2020). There is no shortage of evidence to suggest that this issue was being played out in communities across the country. As one report has noted:

> As a topic of conversation, COVID-19 is the water cooler *wunderkind*.
>
> But not everyone is joining the chat. Multilingual crisis communication has emerged as a global challenge. Australia's migrants, refugees, the aged and those with low literacy levels have struggled to process public health information about the virus (Piller 2020, paras 9–10).

Again, a perfect storm that places particular individuals at a greater risk of contracting the disease.

It is probably fair to say that much of our discussion around these social patterns of disease in the COVID-19 pandemic has focused on individuals and their outcomes. The daily case number reporting and mortality figures that became part of the daily media cycle during the past couple of years are testament to this. But there is another level of interest and concern to consider when we begin to unpack the pandemic's journey through Australian society. For many observers, the connections between socio-economic features of the population and the disease outbreak extend beyond individuals, groups and institutions to become expressed within the social geography of our cities. That is, in urban health geographies. Understanding the health geography of the pandemic is both an exercise in understanding the social determinates of health and illness and a realisation that the impacts of the disease and its responses have furthered already sharp social divides in society. It also draws attention to the fact that when these are played out in space, the often-negative impacts are multiplied, leaving long-lasting legacies.

3.2 The Forgotten Vulnerable: Lost in Space

It is a well-known fact that cities bear the stamp of the broader social structure existing within society. Way back in the work of scholars such Engels (1892) and physician Snow (1855), people began to observe the accumulated social structure being played out in space. For Engels, it was about understanding the great cities and considering how "consequences of our social state…manifest themselves here undisguised" (Engels 1892, p. 65), while Snow's seminal foray into health geography showed us the public health links between the social structure of the city and the outbreaks of cholera. Snow moved our thinking from believing disease was transmitted and spread by 'bad air' or 'bad smells' from rotting organic matter to an understanding of how living conditions that included access to a particular water pump site was associated with spikes in the cases of Cholera. Modern cities are no different in that they represent via distinct social mosaics the social characteristics of the people who live, work and play in geographically separate places and where, if one was to dig deep enough, a range of issues, including health outcomes, are overlayed on social space.

Given this, it is no surprise that the patterns that have become a part of the social determinants' explanation of COVID-19 are also played out in distinct spatial ways within our cities. This link between socio-economic structure and health outcomes in terms of COVID-19 and space can be best illustrated in the Australian context by considering the two cities that bore the brunt of the outbreaks—Melbourne and Sydney.

For residents of the Victorian capital of Melbourne, the arrival of COVID on January 25th 2020, spelt the beginning of the disease in the community. That first case came from a man who travelled from Wuhan, China. As this one case started to multiply, the people of Melbourne began a period of extended lockdowns and strict social distancing and other public health measures introduced by both the federal and Victorian governments. The impacts were widespread and affected even the most mundane of daily rituals. No gyms. No haircuts. No dining out. No visitors to your house. As the Victorian premier Dan Andrews famously stated during one such lockdown period, it is inappropriate to "have your mates around to your home and get on the beers".

But if Melbournites thought they were all in this together, the statistics tell an altogether different story. The city's long-established and unequal urban social hierarchy was displayed with each passing day. If you lived at the wrong address, your odds of contracting the disease were higher than average. The often-quoted phrase "we are all in this together" that was favoured by our politicians should really have been, *we are in this together, except when we are not*. Each new outbreak had a characteristic set of geographically defined hotspots.

Take the outbreak and spread of the Delta variant in August 2021. Coming during a phase when Melbourne was unintentionally vying for the record as the city with the most prolonged COVID lockdown, and just as residents were eyeing a reduction in public health measures, a new outbreak arrived. Since mid-July, Victoria had just celebrated its first day of zero new locally acquired cases. But the doughnut-day, Australian slang for a zero-case day, the euphoria was short-lived when a teacher at an Islamic school in the city's west tested positive. It was the second time that Delta had visited the state. One newspaper headline read *Melbourne plunges back into coronavirus lockdown a week after leaving it* (Miller 2021a). And fall, they did. As a result of the new wave, around 7 million people in Melbourne and surrounding areas of Victoria entered a snap five-day lockdown. Despite this, the cases continued to climb. Although no Local Government Area (LGA) was spared from the Delta variant, the spatial variation in cases was a stark reminder of the uneven nature of city life.

For Melbourne, the Delta outbreak was an outer suburb affair (Fig. 3.1). While there were cases elsewhere, the leafy inner-city regions were impacted to a lesser extent. In contrast, areas on or towards the fringes became the city's emerging COVID hotspots. The sprawling LGA of Hume, located to the northwest of the CBD, topped the list of official cases, recording a rate of 690 per 10,000 population during the period. They became the centre of Melbourne's Delta outbreak and had the locals worried (Henriques-Gomes 2021). Other places such as the LGAs of Casey and Whittlesea were also identified as among the city's pandemic breeding grounds. These places featured in the government's daily briefings as though they were no-go zones for the rest of the Melbourne population. They became a no-go for residents of other states whose governments added them to their lists of 'red zones'.

Melbourne's COVID-19 hotspots also recorded substantial peaks that arrived so much earlier than the peak cases recorded in those communities with much lower numbers (Fig. 3.2). The figure shows the data on new COVID-19 cases displayed as a seven-day rolling average. The explosion of patients in the city's hotspot locations is clear. The evolution of the Delta variant in Hume is a stand-out, with the initial exponential growth and eventual peak coming significantly earlier than in other

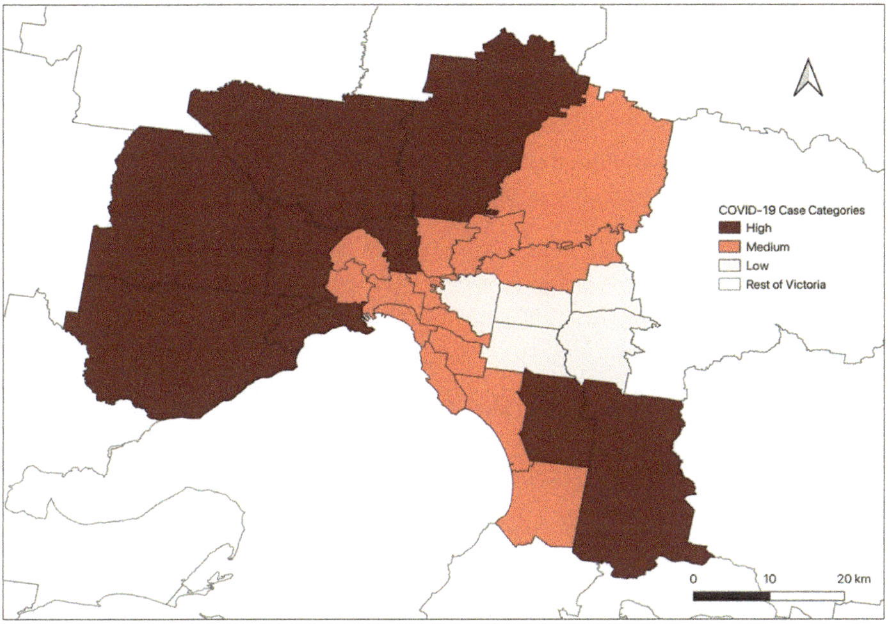

Fig. 3.1 COVID-19 cases, Delta Outbreak, Melbourne Local Government Areas (Authors' calculations from COVID-19 in Australia https://www.covid19data.com.au/)

places. Other hotspot locations recorded high early peaks, but the rise in numbers was later (by two or three weeks), reflecting the later on-set of the initial exponential growth in these places. Although not shown in Fig. 3.2, the other pertinent fact illustrated by the seven-day averages is that the high number of cases lingered for longer in the hotspot locations. Hume still recorded significant numbers of new cases over a month later.

At this stage of the pandemic, it is fair to say that most of us had a reasonably clear idea why the virus was out of control in some areas and not others. We didn't need to read the mountains of social determinants of health literature to understand what was going on. Many of us would have understood that entrenched structural inequality and disadvantage meant that some people were less able to hide from the virus because of their work arrangements or their residential living arrangements. So, just as the social determinants of health literature had warned and the health geographers had written about previous disease outbreaks, Melbourne's spatial COVID-19 patterns were in lock-step with social and economic disadvantage patterns. Local communities that reflected specific characteristics were more likely to witness large disease outbreaks (Table 3.1).

The Melbourne locations making up the high infection rate category had several social dimensions of vulnerability. They tended to have lower levels of income, higher proportions of single parent families, higher unemployment rates and more significant proportions of people with poorer education. This group of locations was also

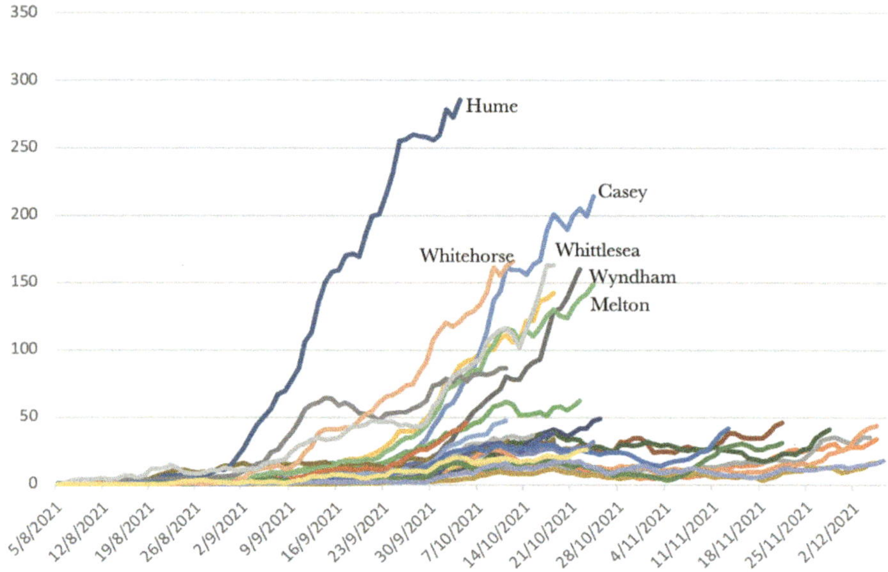

Fig. 3.2 Seven-day moving average and time to peak case numbers Melbourne LGAs, Delta wave (Authors' calculations from COVID-19 in Australia https://www.covid19data.com.au/)

differentiated by having higher levels of people with poor English skills and people who are recent-migrant to Australia. The incidence of household crowding was also higher among this group. In terms of employment patterns, the localities had a significant concentration of blue-collar workers and large proportions of people employed as personal service workers. Finally, when compared to the other categories, the group of local government areas with higher rates of infections recorded a minor reduction on average in workplace mobility, suggesting they were less able to work from home (see Chap. 4).

Unlike Melbourne, where the delta outbreak came on the back of a wide range of existing lockdown measures, Sydney's Delta story began when the people of New South Wales and Sydney had been enjoying a period of COVID-zero. They were enjoying all the singing and dancing and other social benefits of living within their COVID-19 means. As a reporter from the BBC put it

> coasting along almost blissfully detached from the global pandemic. It had achieved a "Covid normal" where people could visit restaurants and nightclubs and join crowds at festivals and theatres (Mao 2021, para 1).

But then the wheels fell off. It started as one case, but within ten days, it had exploded, and Greater Sydney was placed in lockdown. It began with the infection of a limousine driver from the suburb of Bondi and soon became an identified cluster. This cluster soon spread to Western Sydney after a super-spreader event (a birthday party attended by 40 people) and then continued to blow up from there.

Table 3.1 Covid infection rate and selected population characteristics, Melbourne Local Government areas, Infection rate categories (Authors' calculations from Australian Bureau of Statistics 2016 Census data (www.abs.gov.au) and Google Mobility data (www.google.com/covid19/mobility/))

	Covid infection rate category		
	High	Medium	Low
Rate of Covid infections (per 10,000)	470.4	299.4	133.3
Socio-economic status			
Percentage of low-income households	22.8	18.9	15.6
Percentage of households living in public housing tenants	1.3	3.0	1.6
Percentage of households that are single parent families	16.7	13.5	14.0
Unemployment rate	7.9	6.2	6.2
Percentage of population with low formal education	9.9	6.3	6.7
Demography			
Percentage of the population born overseas	41.9	48.1	38.7
Percentage of population who are recently arrived migrants	15.5	12.4	13.5
Percentage of population with poor English skills	14.9	10.8	13.1
Percentage indigenous population	0.95	0.47	0.44
Percentage of dwellings classified as requiring extra bedrooms (crowding)	4.6	3.8	3.0
Employment			
Percentage of part-time employees	31.7	32.5	34.5
Percentage of blue-collar workers	20.5	8.6	9.9
Percentage of sales workers	10.5	9.2	10.0
Percentage of community personal services workers	11.4	9.6	9.5
Percentage change of travel to workplace	−20.4	−32.7	−35.2

The Sydney geography of COVID-19 differed from that of Melbourne. If Melbourne's outbreak was an outer suburb affair, the concentration of Sydney's Delta outbreak, once it spread from its initial point of Bondi, became western suburbs show. A group of Local Government Areas, including Blacktown, Canterbury-Bankstown, Cumberland and Liverpool, became the city's COVID-19 red zones (Fig. 3.3). All of these LGAs saw early exponential growth and, indeed, in the case of Canterbury-Bankstown and Cumberland Shire, saw significant peaks with long tails (Fig. 3.4).

Like Melbourne, Sydney's COVID-19 hotspots could be differentiated on more than just their disease caseload. The Sydney hotspots bore the familiar social and economic signature of outbreaks elsewhere. The locales were home to significant concentrations of disadvantage, low incomes, public housing residents, disadvantaged families, joblessness and poor education. The Sydney hotspots were about migrant communities with high proportions of recent migrants, people born overseas and people with poor English skills. Household crowding was also an issue in

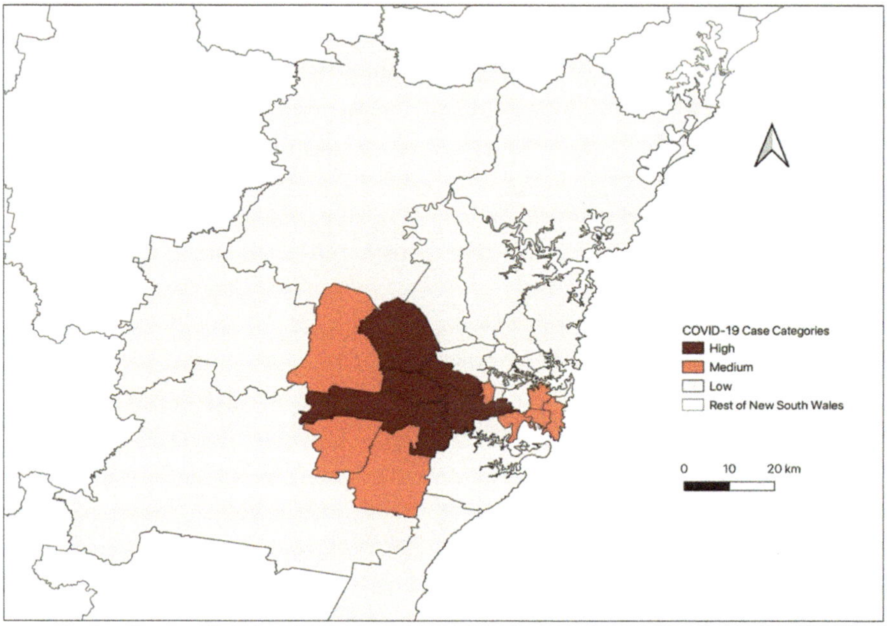

Fig. 3.3 COVID-19 cases, Delta Outbreak, Sydney Local Government Areas (Authors' calculations from COVID-19 in Australia https://www.covid19data.com.au/)

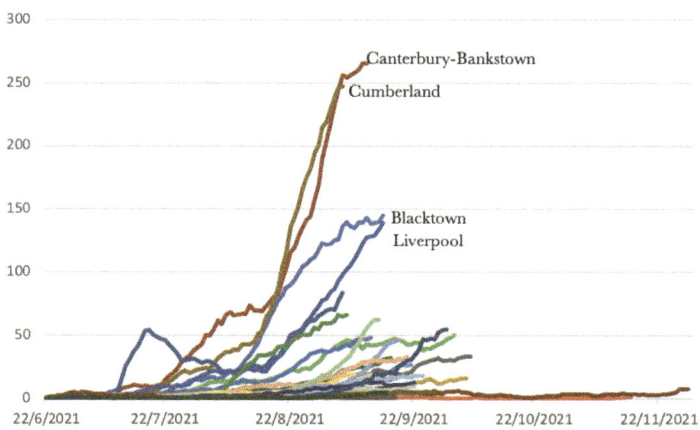

Fig. 3.4 Seven-day moving average and time to peak case numbers Sydney LGAs, Delta wave (Authors' calculations from COVID-19 in Australia https://www.covid19data.com.au/)

these localities. And in employment terms, the hotspot localities had higher concentrations of blue-collar workers and recorded a smaller decline in workplace-related mobility than the LGAs with lower infection levels (Table 3.2).

Across both cities, what the socio-economic patterns of COVID-19 hotspots showed was that a striking social and spatial divergence exists in vulnerability to COVID-19 that runs counter to many of the attempts to simplify the disease risk as one to do with close-contact, connectivity or simple density arguments (Nathan 2020). We see that it is about housing, labour markets and deprivation. Areas harder hit by the pandemic are more disadvantaged, with residents who are more likely to face exposure at work—they are less likely to be able to work from home—or at home due to crowded living arrangements. In addition, cultural practices that encourage large family gatherings, most likely within migrant families, also up the risk. And then, there is the issue of language and risk communication. As we have seen often repeated in the media, communities with concentrations of residents with

Table 3.2 Covid infection rate and selected population characteristics, Sydney Local Government areas (Author's calculations from Australian Bureau of Statistics 2016 Census data (www.abs.gov.au) and Google Mobility data (www.google.com/covid19/mobility/))

	Covid infection rate category		
	High	Medium	Low
Rate of Covid infections (per 10,000)	305.3	132.1	45.0
Socio-economic status			
Percentage of low-income households	31.1	21.5	15.6
Percentage of households living in public housing tenants	6.9	5.6	2.3
Percentage of households that are single parent families	14.2	12.2	8.6
Unemployment rate	8.5	6.1	4.4
Percentage of population with low formal education	11.5	7.8	5.4
Demography			
Percentage of the population born overseas	46.2	34.8	23.5
Percentage of population who are recently arrived migrants	16.9	15.3	12
Percentage of population with poor English skills	23.1	13.5	11
Percentage indigenous population	5.5	7.5	6.7
Percentage of dwellings classified as requiring extra bedrooms (crowding)	3.6	1.9	0.9
Employment			
Percentage of part-time employees	28.8	26.7	28.8
Percentage of blue-collar workers	22.2	17.5	8.8
Percentage of sales workers	9.6	9.4	8.6
Percentage of community personal services workers	10.1	9.9	9
Percentage change of travel to workplace	−34.9	−37.4	−43.1

poor English skills are less likely to heed warnings or fully comprehend the situation. The spatial patterns represent the perfect storm that brings together the structural inequalities that have plagued Australian cities such as Melbourne and Sydney and pandemic risk.

3.3 It Wasn't Only Case Numbers That Were Haunting the Forgotten Vulnerable

It is no secret that the uneven health impacts of the COVID-19 pandemic reached much further than simple case numbers. The unequal geography of cases is also reflected in the loss of life as different social groups and communities faced much higher mortality odds. Relative to other countries, Australia did not witness high mortality rates early in the pandemic, however, with the arrival of the Omicron variant late in 2021, mortality rates began to spike daily. Although the daily press conferences only mentioned totals, a deep dive into the data shows just how some communities were hit.

Melbourne's first case of the Omicron was reported in early December 2021, and like other places, community transmission of the variant soon exploded exponentially. The infection patterns of this variant were much more widespread than in the previous outbreaks. However, as with the Delta and earlier variants, the episode hit disadvantaged and migrant communities much harder (Table 3.3). They had proportionally more cases and deaths. Grouping the Local Government Areas by high, medium and low deaths per 100,000 people, we can see that the LGAs with higher mortality rates were more likely to be characterised as lower-income communities, with higher proportions of disadvantaged families, higher levels of unemployment and lower education levels. They were also home to large populations of overseas-born residents, including those who had recently arrived in Australia and those with poor English skills. Household crowding was also an issue in these locations. On the employment front, these locations had higher numbers of people working in blue-collar occupations. The forgotten vulnerable was in the firing line again.

3.4 Singling Out the Forgotten Vulnerable

The perfect storm between the negative health impacts of patterns of disadvantage and marginalisation also exposed an ugly consequence of the social and geographical patterns we have witnessed. Early in the pandemic, there were stories about racist incidents around the Chinese community within Australia and elsewhere. One survey found that almost one-third of Australians blamed the Chinese community for the emergence of COVID-19 (Razik 2020). Worse still, in early 2020, we read reports in the media of a database reporting 178 racist attacks on Asian-Australians during a

Table 3.3 Covid death rate and selected population characteristics, Melbourne Local Government areas (Authors' calculations from Australian Bureau of Statistics 2016 Census data (www.abs.gov.au) and (Convery and Nicholas, 2022))

	Covid deaths category		
	High	Medium	Low
Rate of Covid deaths (per 100,000)	29.6	20.6	10.9
Socio-economic status			
Percentage of low-income households	26.2	21.0	18.5
Percentage of households living in public housing tenants	2.59	2.11	2.52
Percentage of households that are single parent families	16.4	14.9	13.7
Unemployment rate	8.5	6.6	6.2
Percentage of population with low formal education	10.4	8.4	6.0
Demography			
Percentage of the population born overseas	50.9	39.5	37.9
Percentage of population who are recently arrived migrants	17.7	12.4	12.8
Percentage of population with poor English skills	22.5	13.6	9.9
Percentage indigenous population	0.77	0.83	0.79
Percentage of dwellings classified as requiring extra bedrooms (crowding)	6.7	3.7	3.8
Employment			
Percentage of part-time employees	33.7	33.6	31.6
Percentage of blue-collar workers	17.9	13.2	10.2
Percentage of sales workers	9.4	9.9	9.4
Percentage of community personal services workers	10.2	10.1	9.9

two weeks period. Given the Chinese population's concentration and the incidence of disease, these attacks had a very urban focus. However, as the pandemic rolled on, the focus broadened, and the search for COVID-19 villains became a daily sport in the media and through the actions of politicians. Those identified as patient zero in a new outbreak or the residents of communities with the highest disease burden were blamed for not abiding by the public health directives (Miller 2021b). We heard about the barbecue man, the border jumpers, the limo driver, the removalists and the pizza man. These were widely used nicknames in the press and other media outlets for people who have found themselves thrown into the national headlines and blamed for spreading coronavirus. Very early in the pandemic, three young women from Brisbane lied about visiting a Coronavirus hot spot to avoid hotel quarantine. Two of them contracted COVID, and they were dubbed "the border jumpers" with their faces plastered on the front of Queensland's biggest newspaper with the headline "Enemies of the State". While some of these people had breached public health orders, others had not. But this didn't stop the national blame game and outrage from taking off. Some of the public took offence at the media's shaming of these "COVID villains"

Fig. 3.5 Social media response to the shaming of COVID villains (Miller 2021b)

(Fig. 3.5), pointing out that in many cases, people did not breach any public health requirements but were simply in the wrong place at the wrong time.

COVID-19 victimisation didn't stop at individuals but saw whole communities in the firing line. In the initial stages of the pandemic, the people of Melbourne witnessed this first-hand when the government decided to shut down entire public housing blocks for fear of a widespread outbreak. Although this was often seen as a policy focusing on the dangers of high-density living (see Chap. 1), others saw it differently. Sociologist Zevallos (2020) saw the Victorian government's actions as effectively criminalising the poor by using public health orders to enable a police-enforced lockdown of the public housing blocks. This was, for Zevallos, the beginning of the moral panic being shaped by the pandemic:

> COVID-19 has dusted up a predictable moral panic. Poor people, migrants, refugees and Aboriginal people are, on the one hand, positioned as vulnerable and in need of additional state intervention because they are at higher risk of infection. On the other hand, these disadvantaged groups are simultaneously characterised as "folk devils" who are responsible

> for spreading the disease. Either way, the moral panic justifies the increased policing of underprivileged groups (Zevallos 2020, para 11).

The unequal treatment of whole communities didn't stop with locking down public housing estates. Communities in the firing line also saw inequities in how justice was being served. Although there was no official data, there were allegations that police in Melbourne were over-policing disadvantaged and migrant communities:

> In some regions of Melbourne, there are more police and more intense levels of police, and those areas are more likely to have more intense COVID-19 public health restriction policing by itself (Blakkarly 2020, para 4).

In New South Wales, police were criticised for what appeared to be a practice of unevenly issuing social distancing fines. Fines given out for breaches of COVID-19 restrictions were much higher in the marginalised communities in the west and south-west of Sydney. Not by accident, these communities also have a higher proportion of culturally and linguistically diverse residents and are the same places we have identified in Fig. 3.3 and Table 3.3 above. We read that,

> One-third of infringements were issued in highly multicultural suburbs, such as South-West Sydney, even though this region had a relatively low transmission rate. Yet the top four local government areas (Waverley, Northern Beaches, Inner West, and Randwick) had one quarter (23%) of locally acquired COVID-19 cases, but only one infringement. These are majority Anglo, White and affluent suburbs (Zevallos 2020, para 17).

Data for the 2021 Delta wave in New South Wales shows the striking disparities in the issuing of infringements, with the most disadvantaged post-codes receiving, on average, almost 11 fines per 1,000 people compared to a measly 4 for more affluent post-codes (Fig. 3.6) (Rachwani and Evershed 2022).

During later waves, entire communities were blamed for not following public health orders. In New South Wales, the government introduced a two-tier system of lockdowns targeting Sydney locations with higher case numbers and implementing much tighter restrictions. Added to this, the government decided that these communities needed some stern guidance, so they arranged to send in the police and Army to make sure members of these selected lockdown communities "did as they were told." This was despite social media posts showing that residents of other localities were flouting the strict rules.

The impact was to further isolate and alienate these vulnerable communities. In one media report at the time, Rachwani (2021) proclaimed "Fear and loathing in western Sydney: how NSW's COVID response failed migrant communities" and argued

> Tough policing, constant rule changes and health messaging that isn't cutting through have created a perfect storm of misinformation and paranoia in Sydney's most diverse areas (Rachwani 2021, para 1).

At the same time, we read about "Troops enforcing western Sydney lockdown" (Rachwani and Allam 2021) and how the government was

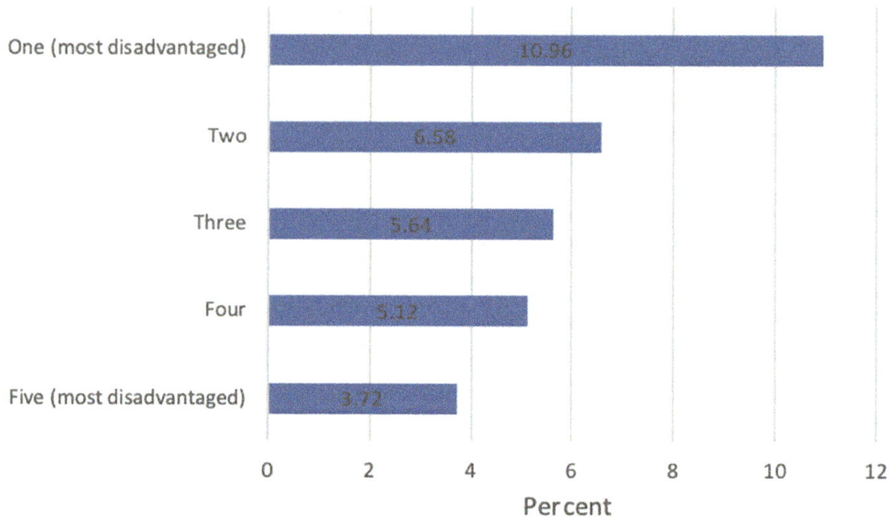

Fig. 3.6 COVID-19 fines by the socio-economic status of postcode Source (Rachwani and Evershed 2022)

> ... choosing to use force and fear by giving police additional powers and rolling in army troops to communities is disturbing. This will only create more significant anxiety and uncertainty among our families and friends (para 9).

The author of this article quoting a resident, summed the situation up nicely, stating:

> Their deployment is such a statement about the nature of the problem, and the problem is us, the people who live in western Sydney. They're saying the problem isn't the vaccine rollout or their failure to support people, the problem is our compliance (para 14).

A master class in blaming the victim!

3.5 The Continued Disadvantage of the Hidden Vulnerable

In lots of ways, it is no surprise that the highest impact of COVID-19 has been felt mainly in some of the most disadvantaged communities within our cities. People living in disadvantaged communities have had notoriously poorer health outcomes across various diseases and health issues. The COVID-19 outbreak has accentuated these disadvantages that have been entrenched in our cities for decades. And it is in these areas, that the fallout from the pandemic will keep on giving, and not in a good way.

The social and economic impacts of the pandemic are most likely to be felt in the most disadvantaged communities long after COVID-19 has left the front pages of our newspapers. The labour market impacts (see Chap. 4) will likely impact underprivileged communities, resulting in overwhelming economic scarring. And the rough treatment of some communities, in terms of over-policing and victim-blaming, will heighten racial and cultural divides in many places.

But the impacts in terms of the continued disadvantage of the hidden vulnerable will also be witnessed in more subtle and potentially long-lasting terms. Being a child in a disadvantaged area and a low-income family or household can be hazardous to your future. At the very least, you may be starting behind your peers living elsewhere. The impacts of COVID-19 on disadvantaged families and communities will make this worse. Think about what might happen if you grow up in a family where the role models (i.e. parents or older siblings) have suffered from significant periods of joblessness. Jobless families existed before the pandemic, but the economic hit caused by the public health measures exacerbated the problem in many areas. To get an idea of the potential size of the problem, we can look at the Australian Bureau of Statistics labour force estimates of families produced from data collected in the monthly labour force survey (LFS). Considering the October 2021 (Australian Bureau of Statistics 2021) release, the official data suggests

> In June 2021, there were 93,500 jobless couple families with dependants (including children under 15 and dependant students aged 15-24 years), down from 133,000 in June 2020. 80,000 (85.4%) of these families had children under 15. An estimated 162,000 children aged 0-14 years were in jobless couple families … (para 41).
>
> There were 190,000 jobless one parent families with dependants in June 2021, about one third (29.1%) of all one parent families with dependants. 90.2% of these jobless one parent families had children under 15. This equated to an estimated 315,000 children aged 0-14 years in these families (para 42).

We can see from the data that even in earlier years (2018), the number of children living in jobless families (especially those aged 0–14 years old) was significant, illustrating that the problems associated with jobless families are not just a COVID phenomenon (Fig. 3.7), but it has undoubtedly been exacerbated by it.

But why is this a problem? Put simply; individuals are significantly more likely to be disadvantaged in the labour market (face unemployment) if their parents had not been in the paid workforce when they were children—even taking into account all of their other characteristics (that is, level of education, age, sex, ethnic status, etc.). So, for a child, living in a jobless family is not good, and in the case of the current employment crisis, the adverse employment implications for specific groups will likely be long-lasting. Like other social problems, the health crisis and the accompanying economic hit have made these issues far more visible. And they are worse if these families are concentrated in the same communities.

Those most impacted by COVID-19 and living in jobless families and disadvantaged communities are likely to face extreme financial pressures, which has been shown to affect educational attainment and other adverse social outcomes. This double whammy leads to further disadvantages when young people move or attempt to move into the labour market. Poor education attainment and a lack of exposure to employment role models equal poor life chances.

In the Australian context, we can read some recent findings in an academic paper by Bubonya and Cobb-Clark (2021), where the authors point out that:

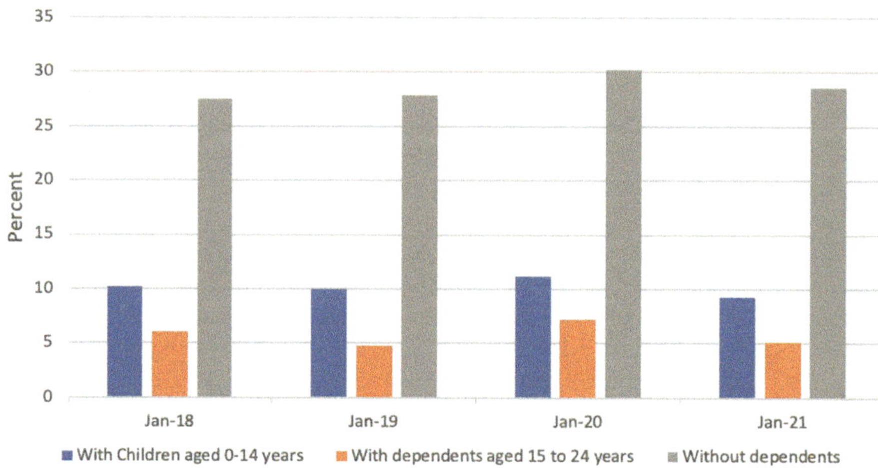

Fig. 3.7 Jobless families by the age of youngest dependent, June 2018 to June 2021 (Australian Bureau of Statistics 2021)

> intergenerational persistence in economic disadvantage can broadly be understood through an education lens. The most critical mechanism linking welfare across generations is the failure to complete high school (p. 14)

and

> Disadvantage is also being perpetuated from one generation to the next through the parenting that young people receive. The issue is not the style parents adopt (i.e., warmth and monitoring) but rather the relative lack of support—particularly financial support—that welfare-reliant parents provide. Without adequate parental support, young people's ability to successfully transition from education to employment may be constrained (p. 14).

Added to this is the burden imposed by the transition to homeschooling during the pandemic, which may have impacted children in disadvantaged families (those most at risk of seeing some form of negative intergenerational impacts) who may struggle to catch up with their educational levels. Some of this will be an outcome of long-term academic disengagement, whereby

> vulnerable young people already have a precarious relationship with education, there is a possibility that they may further disengage from learning if the curriculum content is only provided online (Drane et al. 2020, p. 6)

But the educational burden will also be perpetuated by the diminished digital inclusion suffered by those among the most vulnerable. Despite the Australian government's National Broadband Network rollout, disadvantaged households and communities were likely to face greater difficulty due to poor internet access. During the working or studying from home orders, we heard stories from support agencies about the impact that poor internet access was having:

> COVID-19 leads to more people trying to access government services online, people being forced to work from home, and people needing online access to health and education. It is also our primary way of connecting to family, friends and the outside world (para 6),

> But as this research demonstrates, this is simply out of reach for many people in our community and will only lead to more significant economic and social dislocation (para 7) (Visentin 2020).

It appears then that when we consider the most vulnerable communities and populations in the COVID-19 story, there has been significant collateral damage in terms of both the burden of the disease and the potential long-run consequences. The COVID-19 pandemic has made the uneven nature of our cities and communities even more apparent, which the public health measures and economic impacts have potentially made worse. When we eventually move on from the pandemic stage of the disease to a more endemic stage, these same communities will likely continue to be the hardest hit due to their social and economic characteristics. In short, the meaning of long-COVID may be very different for these individuals and the communities they live in.

References

Australian Academy of the Humanities (2020) Fighting COVID-19: it's time to break down the language barriers. https://humanities.org.au/power-of-the-humanities/fighting-covid-19-its-time-to-break-down-the-language-barriers/

Australian Bureau of Statistics (2021) Labour force status of families. https://www.abs.gov.au/statistics/labour/employment-and-unemployment/labour-force-status-families/latest-release#methodology/

Australian Institute of Health and Welfare (2021) The first year of COVID-19 in Australia: direct and indirect health effects. Australian Government. https://www.aihw.gov.au/reports/burden-of-disease/the-first-year-of-covid-19-in-australia/summary

Blakkarly J (2020) Concerns police using coronavirus powers to target marginalised communities in Australia, SBS News, https://www.sbs.com.au/news/article/concerns-police-using-coronavirus-powers-to-target-marginalised-communities-in-australia/p36wv2d1y

Bubonya M, Cobb-Clark DA (2021) Pathways of disadvantage: unpacking the intergenerational correlation in welfare. Econ Educ Rev 80:102066

Convery S, Nicholas J (2022) 'Disease of disadvantage': Melbourne's lower socioeconomic areas suffer most Covid deaths amid Omicron. The Guardian. https://www.theguardian.com/australia-news/2022/feb/25/disease-of-disadvantage-melbournes-lower-socioeconomic-areas-suffer-most-covid-deaths-amid-omicron

Drane C, Vernon L, O'Shea S (2020) The impact of 'learning at home'on the educational outcomes of vulnerable children in Australia during the COVID-19 pandemic. Literature review Prepared by the National Centre for Student Equity in Higher Education. Curtin University, Australia

Engels F (1892) The condition of the working-class in England in 1844 with preface written in 1892. Swan Sonnenschein & Co, London

Henriques-Gomes L (2021) 'They're terrified': how Melbourne's City of Hume became the centre of Victoria's Delta outbreak'. The Guardian Australia. https://www.theguardian.com/australia-news/2021/sep/18/theyre-terrified-how-melbournes-city-of-hume-became-the-centre-of-victorias-delta-outbreak

Herath S, Bentley R (2018a) Crowding, housing and health: an exploratory study of Australian cities. Paper presented at the State of Australian Cities National Conference

Herath S, Bentley R (2018b) Overcrowded housing looms as a challenge for our cities. The Conversation. https://theconversation.com/overcrowded-housing-looms-as-a-challenge-for-our-cities-96110

Mao F (2021) Covid: how delta exposed Australia's pandemic weaknesses. BBC News. https://www.bbc.com/news/world-australia-57647413

Miller M (2021a) Melbourne plunges back into coronavirus lockdown a week after leaving it. The Washington Post. https://www.washingtonpost.com/world/asia_pacific/delta-australia-melbourne-lockdown/2021a/08/05/c53cbe78-f5b6-11eb-a636-18cac59a98dc_story.html

Miller M (2021b) A search for villains': as Australia's outbreak grows, so does covid shaming. The Washington Post. https://www.washingtonpost.com/world/2021b/07/23/covid-australia-lockdown-shaming/

Nathan M (2020) The city and the virus. Urban Stud. 00420980211058383

Patel J, Nielsen F, Badiani A, Assi S, Unadkat V, Patel B, Wardle H (2020) Poverty, inequality and COVID-19: the forgotten vulnerable. Public Health 183:110

Piller I (2020) Lost in translation: COVID-19 leaves migrants behind. The Lighthouse. https://lighthouse.mq.edu.au/article/october-2020/lost-in-translation-covid-19-leaves-migrants-behind

Rachwani M (2021) Fear and loathing in western Sydney: how NSW's Covid response failed migrant communities. The Guardian. https://www.theguardian.com/australia-news/2021/aug/13/fear-and-loathing-in-western-sydney-how-nsws-covid-response-failed-migrant-communities

Rachwani M, Allam L (2021) Troops enforcing western Sydney lockdown will alienate community, advocates warn. The Guardian. https://www.theguardian.com/australia-news/2021/jul/30/troops-enforcing-western-sydney-lockdown-will-alienate-community-advocates-warn

Rachwani M, Evershed N (2022) Incredible imbalance: NSW Covid fines during Delta higher in disadvantaged suburbs. The Guardian. https://trem.media/australia-news/datablog/2022/feb/10/incredible-imbalance-nsw-covid-fines-during-delta-higher-in-disadvantaged-suburbs

Razik N (2020) Almost one third of Australians blame coronavirus pandemic on Chinese community, new poll finds. SBS News. https://www.sbs.com.au/news/almost-one-third-of-australians-blame-coronavirus-pandemic-on-chinese-community-new-poll-finds/2f11e2b6-2e94-4f73-8892-93e63e6e8656

Snow J (1855) On the mode of communication of cholera: John Churchill

Visentin L (2020) 'Digital divide': lower income households struggle with internet access. The Sydney Morning Herald. https://www.smh.com.au/national/nsw/digital-divide-lower-income-households-struggle-with-internet-access-20200416-p54kfm.html

Zevallos Z (2020) Pandemic, race and moral panic. https://othersociologist.com/2020/07/05/pandemic-race-and-moral-panic/

Chapter 4
The Great Job Reshuffle: How COVID-19 Changed Urban Labour Markets

4.1 Introduction

As we have already seen, the COVID-19 pandemic has impacted almost every part of life and has left very few people or communities untouched. Although primarily a health emergency, it quickly developed into a parallel economic crisis impacting the livelihoods of individuals, families and communities. As the economic 'engine rooms' of the nation, large cities were the first to battle the economic headwinds created by what would become the COVID-19 induced recession. Other urban regions and large regional centres were soon embroiled in the slowdown as government-mandated shutdowns or stay at home orders were introduced, international borders were closed, and all but essential services ground to a halt or significantly transitioned. Apparently, everyone would be staying at home hiding under their collective doonas.

The reshuffling of urban labour markets resulted in several expected outcomes, like increases in joblessness which raised questions about the potential long-term harm through social scaring and financial stress, and as we saw in the last chapter, the impacts of spatially concentrated 'clusters' of unemployed can have on the broader community social fabric. But the reshuffling also raised questions about the shape of urban labour markets as some workers pivoted to working at home, impacting commuting patterns and the viability of Central Business District-based businesses reliant on a steady flow of daily customers for their survival.

The Australian experience was not in itself unique. We saw in daily news reports stories of entire cities and countries grinding to a halt, CBDs deserted, turned into instant ghost towns. Data showing worrying trends—COVID-19 cases and deaths going up (often dramatically) and economic indicators retreating.

Dealing with COVID-19 was like using drugs to treat a disease; the treatment slows down the progress of the illness but may negatively impact other bodily systems. So too for the case of dealing with COVID. Shutting down the economy made the disease less likely to spread, but in the process, other societal systems were damaged. Economists might refer to these as negative externalities or spillover effects. Whatever you want to call them, they hit fast and hard!

S. Baum et al., *Pandemic Cities*, Cities Research Series,
https://doi.org/10.1007/978-981-19-5884-7_4

Once the pandemic became widespread, alarm about the economic consequences started to emerge globally, though they lagged by several weeks. By mid-February 2020, the Organisation for Economic Co-operation and Development (OECD) began sounding warning bells about the possibility of the global economy entering a severe recession. OECD estimates put the initial contraction in output for various economies between 20 and 25%, with longer-term impacts wiping off around two percentage points of annual GDP growth for every month of strict lockdowns (OECD 2020). This was a bad set of predictions. They warned that as economies began to contract, this would mean significant increases in joblessness and further economic and social pain on top of the stress of a health pandemic. No wonder politicians and others were scrambling to introduce measures to offset the negative economic impacts and attempt to shelter as many workers as possible from the predicted upswing in unemployment numbers.

4.2 The Australian Economy and the Onset of COVID-19

Leading into the COVID-19 pandemic, the Australian economy was already showing signs of pressure. Despite what politicians would want us to believe, the country's economy was already under the pump. The country had endured a sequence of unplanned disasters in the 12 months leading up to the pandemic's start. The lingering effects of an extended period of drought, devastating bushfires and floods. On top of this, labour markets were already feeling the impact of inadequate demand, with unfavourable labour market indicators already heading in undesirable directions.

One indication of the trouble we were facing can be gleaned from data on our economic growth. The standard measure of growth, GDP or Gross Domestic Product, was not looking great. Real per quarter GDP growth was already below the average trend of 3.3%, and looking at the figures, it is clear just how far below historical trends the growth performance had been (Fig. 4.1). Even with an enormous household spending spree and a boom in the mining sector, Australia's real GDP growth was declining significantly below the long-term performance.

Away from GDP, the indicators that cut close to home with many Australians reflect jobs, or more importantly the lack of jobs. Nothing quite gets politicians' hearts racing and palms sweating more than an adverse movement in jobs; after all, being able to get people into jobs is something our elected officials hang their economic management credentials on. But despite the hype that the Australian labour market was in a strong position, like GDP, the figures tell a different story (Fig. 4.2). Two measures of labour wastage—the underutilisation rate and the underemployment rate—were already showing signs of deterioration even before the pandemic hit. The underemployment rate, which accounts for those people working less than full time who want to work extra hours, has seen a gradual increase over the last three decades and has continued to kick up in the post-Global Financial Crisis (GFC) period. Even before the start of the COVID pandemic, the underutilisation rate was higher than in the 2009/10 period.

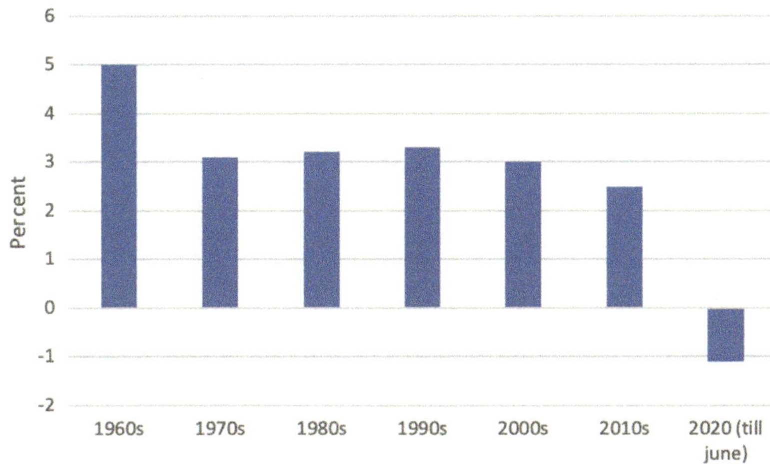

Fig. 4.1 Real GDP growth per quarter, Australia (Mitchell 2020)

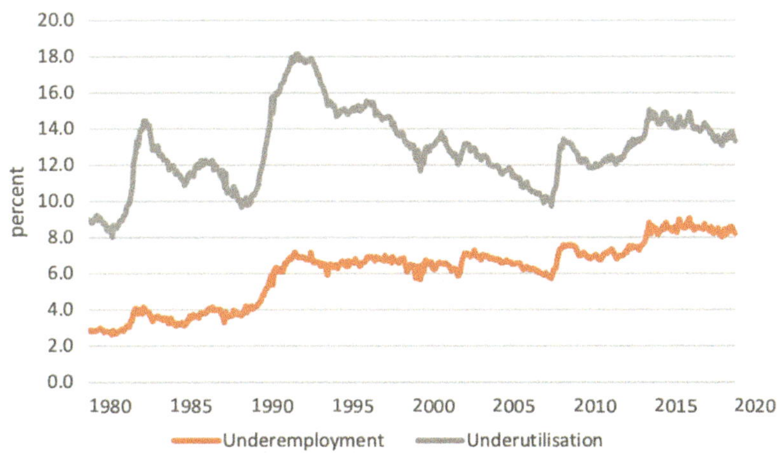

Fig. 4.2 Broad underutilisation and underemployment rate, Australia (Australian Bureau of Statistics 2021)

But it is the overall underutilisation rate that is more alarming. The underutilisation rate, which adds the unemployed to the underemployed rate, provides a broader measure of labour is being wasted. And it is clear that the broad underutilisation rate has tracked the underemployment rate since the GFC period. And we can see that like underemployment, underutilisation was already tracking north in the period prior to the COVID-induced shutdown.

So even before the initial COVID-19 slowdown, the economy in both growth terms and labour market resilience was not in good shape. Not really an ideal place to start from. What the COVID-19 pandemic did was exacerbate these negative trends in ways that not many politicians, economists or laypeople had envisaged. In 2019, economic commentators predicted a period of small economic growth in 2020 (Chalmers 2019). These so-called expert economic commentators predicted that.

> A slightly better 2020 than 2019 but nothing to shoot the lights out (para 8)

and that

> Due to high population growth, a technical recession remains unlikely in Australia (para 50).

Sure, these 'experts' couldn't see the COVID-19 train approaching, but their optimism illustrates how everyone was surprised.

At the onset of the COVID-19 health pandemic and following the warnings by the OECD, there was talk of Australia slipping into its first recession since the early 1990s. Not even during the Global Financial Crisis did the Australian economy officially enter a technical recession, defined as a situation where growth in economic output goes 'negative' for two quarters in a row. Many countries did. But Australia, via good management or good fortune, managed to see off the economic headwinds on that occasion. But not this time.

The concern about a recession and the harm it would cause was palpable. Throughout March 2020, headlines such as *Buckle up for a hell of a bumpy ride* (Creighton 2020) or *Coronavirus has delivered a hammer blow to Australia's economy* (Butler 2020) became commonplace. And then, with the release of the March quarter GDP numbers, it became official. Now the headlines proclaimed *Australia in its first recession in 29 years as March quarter GDP shrinks* (Janda 2020) and *Australia enters first recession in 29 years after blow from bushfires and coronavirus* (Murphy 2020a). The graphics accompanying the headlines showed a worried-looking then federal Treasurer pointing to graphs with the trends heading south (Fig. 4.3).

At the time, the Australian Bureau of Statistics started to release information based on the number of people employed using information from the Australian Taxation Office. Officially, it was called the Weekly Payroll Jobs and Wages in Australia data. Benchmarked to the date when Australia recorded its 100th COVID-19 case, the data became a barometer of the health of the labour market. In the initial weeks following the escalated national shutdowns, the payroll index fell sharply. It had appeared that employment had fallen off a cliff, with the index falling from its benchmark of 100–92.5. In percentage terms, this represented a fall of 7.5% between the 14th of March and the 18th of April. Bad news indeed.

4.3 The Unequal Labour Market Burden of the Pandemic Shutdown

A number of phrases became common during the pandemic. One of the most fallacious was the phrase 'we are all in this together'. Politicians were fond of trotting it

Fig. 4.3 Former Australian federal treasurer Josh Frydenberg presents the nation's latest GDP data showing the economy is in recession (Murphy 2020b)

out whenever they had a difficult announcement to make and wanted to get 'Team Australia' on board. In reality, as we noted in Chap. 3, perhaps the phrase should have 'we are all in this together…except when we are not".

Across the country, but especially in the metropolitan labour markets, fears about job losses were high. Through its JobKeeper package, the Australian Federal Government reduced the potential unemployment burden for many. Throughout the most critical period of the economic downturn, the government was fond of telling us about the success of the JobKeeper program. The Prime Minister apparently even received fan mail from people whose jobs the support packages had saved. Modelling by the Reserve Bank of Australia suggested that.

> one in five employees who received JobKeeper (and, thus, remained employed) would not have remained employed during this period had it not been for the JobKeeper Payment. Given that 3½ million individuals were receiving the payment over the period from April to July 2020, this implies that JobKeeper reduced total employment losses by at least 700,000 over the same period (Bishop and Day 2020, p. 28).

Despite the lauded successes of the government's package, the nation as a whole still witnessed significant increases in unemployment with long lines forming at government Centrelink offices. Official government figures suggested that.

> By April, the number of employed Australians had fallen by 594 300—including 220 500 from full-time employment and 373 800 from part-time work. 828 300 people were unemployed and the underemployment rate had increased to 13.7 per cent (Select Committee on COVID-19 2020, p. 68).

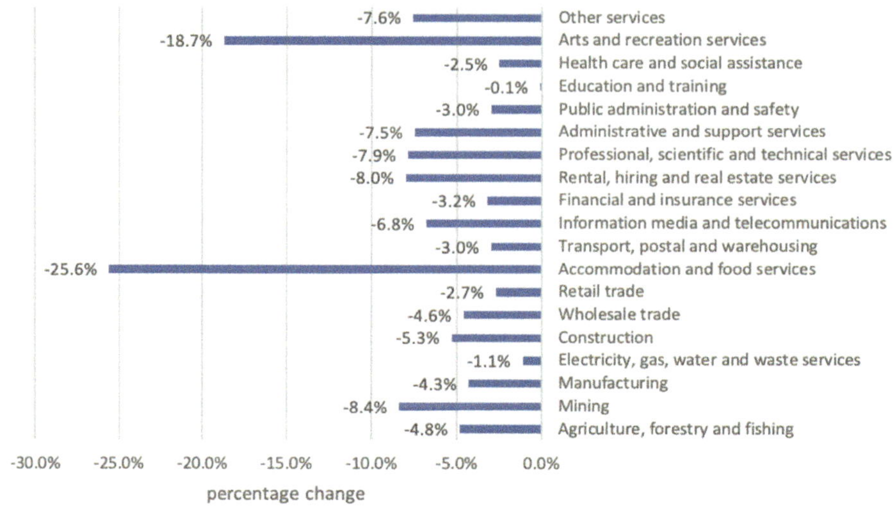

Fig. 4.4 Change in weekly payroll jobs by industry sector (Australian Bureau of Statistics 2020c)

As a result of the significant fall in employment when the pandemic hit, the official unemployment rate surged to 7.5% in July. The unemployment rate continued to grow in June and July, even though some recovery in employment was already underway.

At a broad industry level, the job losses witnessed during these early days were far from evenly distributed (Fig. 4.4). They were heavily concentrated in hospitality and recreation and the arts, sectors that also have the lowest wages.

Reflecting on these early days of the pandemic, former economics editor for the Australian, David Uren, observed that.

> In the three weeks after the first 100 cases of coronavirus were registered, the hospitality industry shed 26 per cent of its workforce, or 240,000 workers, while the recreation and arts industries lost 19 per cent, or 47,000 workers.
>
> Both industries are among the lowest paid, with average weekly incomes of just A$638 in the hospitality sector and A$989 in recreation and the arts (Uren 2020, pp. 2–3).

Figures such as these tell a story about the labour market in aggregate. But there is no such thing as a single labour market, and economic slowdowns are not even-handed (Sandbu 2020). They impact different people differently. Within cities, they are geographically uneven, with the impacts ramped up more in some communities than others, resulting in a spatial pattern representing a patchwork quilt (see Chap. 3). The geography of unemployment changed in the blink of an eye. Whereas once, significant concentrations of unemployment were clustered in some city regions, the new COVID geography of unemployment was different. Sure, there were the usual suspects, but we also witnessed significant increases in joblessness in other localities.

In a report released early in 2020, Scott Baum and William Mitchell pointed out that close to 13% of localities across Australia's metropolitan cities and other large urban regions were in danger of seeing significant increases in joblessness, with another 45% being classified as having ‘medium–high risk’ of increasing unemployment as a result of the COVID induced economic slowdown (Baum and Mitchell 2020). Thanks to the Australian federal government’s fiscal intervention, including its JobKeeper package, the reality was less widespread than the Baum and Mitchell report suggested. However, many cities saw increases in unemployment in places that were not hotspots in the past. The initial increases were seen in large capital cities and other large urban cities and soon spread to smaller places. The geography of unemployment was now completely different than it had been at any time in recent memory.

Take Sydney for example. Often known as the county’s first global or world city (although Melbourne residents might dispute this), it also has a reputation, like other similar cities, for the uneven employment fortunes of its residents and the communities they live in. You could ask Sydney residents which suburbs or communities had concentrations of jobless, and they would be able to, within limits, tell you. But this time, it was different. The impact of the nationwide economic slowdown bought with it a different vibe.

The map of changes in payroll jobs between March 14th 2020 when the country recorded its 100th case of the disease and the end of the year was a sea of red. It isn’t just the usual ‘left behind places’ where migrants, single parents, indigenous people and others face significant labour market disadvantages. However, these did feature heavily as jobs disappeared. Labour markets in some of the city’s most disadvantaged regions, those communities located in the city’s west, saw significant job losses. The performance of the Sydney-South West labour market region, home to the communities such as Fairfield and Liverpool, recorded an initial fall in payroll jobs of close to 8% (in the first month) and a longer-term decline of close to 5%. Worryingly, this was on top of an estimated current unemployment rate of 8% (Australian Bureau of Statistics 2021). What’s more, labour markets such as this appeared less able to benefit from the federal government support packages. In a blog post titled “Using a regional lens reveals the uneven impact of the COVID employment crash” (Baum 2021), it becomes clear that.

> Regions with higher pre-pandemic unemployment saw an initial fall in jobs as the economy headed to a standstill and have, on average, failed to claw back much ground in the interim and at least to a casual glance, actually fallen further behind (para 41).

and that

> More disadvantaged regions in term of pre-pandemic labour market conditions were hammered and the government’s job keeper interventions do not seem to have only made limited inroads in these regions (para 42).

The map of payroll jobs loss also showed other areas in the red (Fig. 4.5). Labour markets comprising localities in the city’s eastern suburbs and northern suburbs, once home to relatively low levels of joblessness, became new jobless hotspots. The labour market region comprising the city and inner-south recorded an initial first-month fall in payroll jobs of 11%, the highest across the entire Sydney region. It recorded a longer-term fall of 6.8%. These significant job losses occurred in an

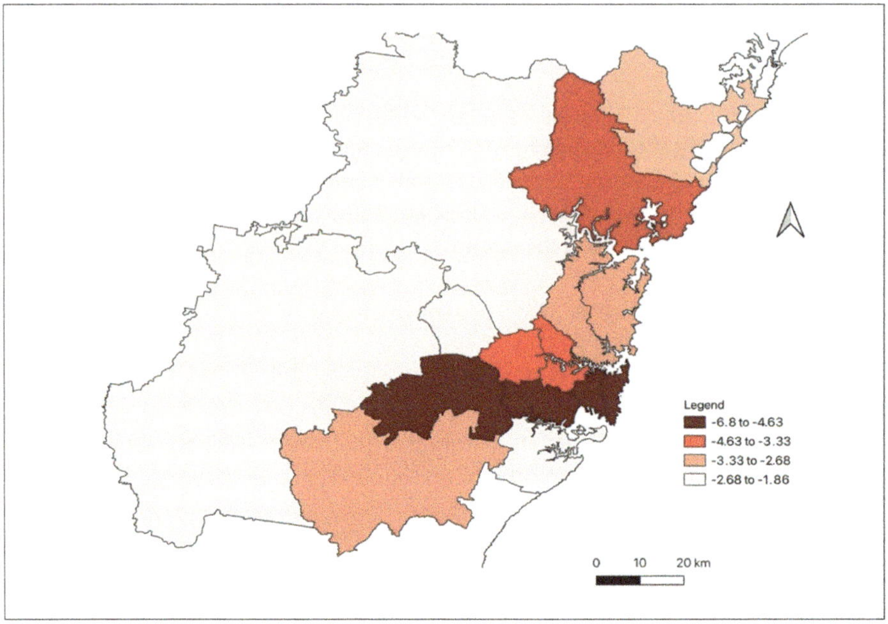

Fig. 4.5 Change in Payroll Jobs, Sydney (Australian Bureau of Statistics 2020d)

area where unemployment rates have been traditionally low—the region recorded an unemployment rate just above 3% before the pandemic.

Other cities were not immune. More or less, the sea of red that had engulfed Sydney was also being played out elsewhere. In Melbourne, we saw significant falls in employment across the inner-city region and inner-south region. The inner-city region continued to witness a significant employment decline through 2020 and into 2021, thanks to the extended lockdown imposed on the residents of Melbourne.

The reshuffled geography of unemployment across cities has significant consequences for much of how we understand the impacts of the pandemic-induced economic shutdown and has reached far beyond simply understanding labour market outcomes. Importantly, it has implications for the types of inequalities and disadvantages we talked about in chapter two.

In terms of these uneven jobless outcomes, the pandemic will keep giving, potentially impacting urban labour markets for some time to come. Despite the view that once the private sector starts firing again, the economy will "roar back to life", it is more likely that the adverse labour market impacts of the pandemic will be with us for a while and that the impacts will be highly uneven. Considering the unemployment and labour underutilisation trends following previous downturns (Figs. 4.6 and 4.7), we can begin to understand the likely lingering impacts of the COVID-induced slowdown. From this data, we learn that when we compare unemployment trends with previous economic downturns (the 1982 recession, the 1991 recession and the

GFC), the labour market will take some time to bounce back once an unemployment peak is reached.

The truth is that when the recovery does get properly underway, gains will be patchy, and there will be winners and losers. Some will move relatively quickly back into employment, while others, often the most disadvantaged, will find it extremely

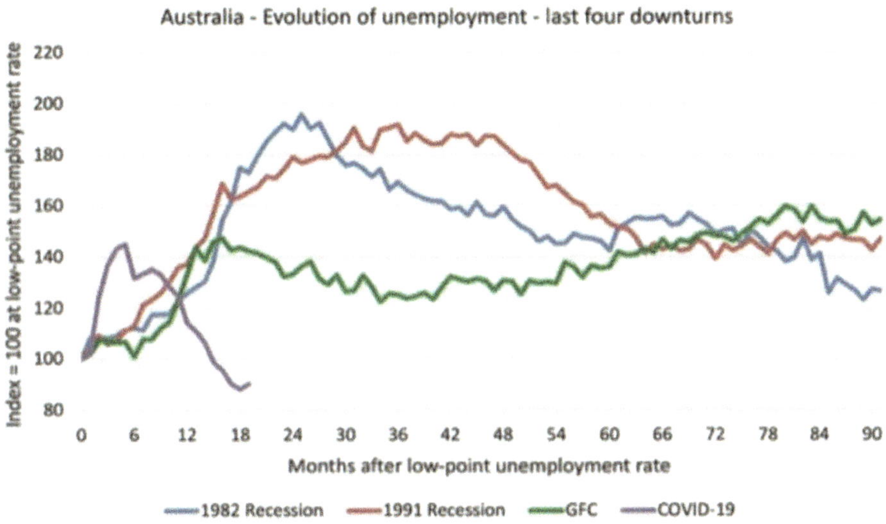

Fig. 4.6 Evolution of unemployment during the last four downturns (Mitchell 2021)

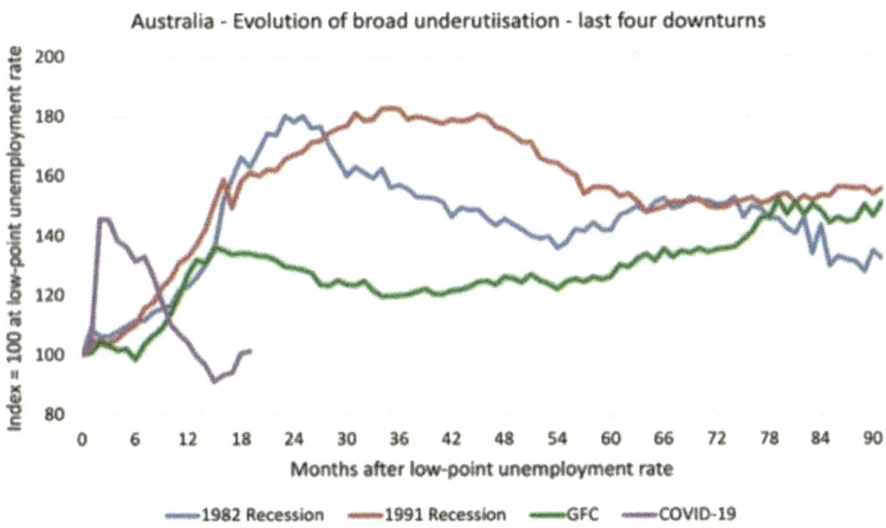

Fig. 4.7 Evolution of underutilisation during the last four downturns (Mitchell 2021)

difficult. And this will have noticeable impacts on the uneven geography of unemployment. This point is well illustrated in the Report from Anglicare (Anglicare 2021) titled *Australia's Job Availability snapshot 2021*, where the authors of the report point out that:

> Most employment statistics assume that each person can compete for every job. We know this isn't true … [and] … In spite all of the changes to the workforce over the past year, the number of people with barriers to work has barely budged. Even with a resurgence in the number of entry-level job vacancies, people with the most significant barriers to employment aren't getting them. They are competing with 27 jobseekers for each one of these roles (para 3).

Any notion that the adverse employment impacts of the COVID-19 economic slowdown will be over once the economy "roars back to life" are not accounting for reality, and all reports that suggest otherwise should be considered with scepticism.

For Australian cities, the uneven regional or spatial labour market context within which recovery will take place will mean that many people and their communities will face ramped up long-term disadvantages, impacting the types of patterns of weakness. For others, the negative impacts of the economic slowdown will be much more transient. How this plays out will depend on the long-term ability of different regional economies to recover from the COVID-19-induced economic downturn and the level of employment resilience in other localities. This will depend on many factors, including its pre-pandemic economic position, its industrial mix, the time and extent of public health-imposed shutdowns and stay at home orders.

Put simply, local economies and labour markets will face uneven paths out of the slowdown, which will have significant impacts on the individuals who live there. Regardless of your individual characteristics (level of education, family background, etc.), the type of regional labour market you live in has a significant impact on your employment opportunities. The fact that different regions will recover at different speeds will mean that out-of-work individuals in poorly performing regions will be significantly disadvantaged in attempting to re-engage with the paid workforce. That seems like a no-brainer but is often overlooked when the policy is considered, and politicians talk about how robust the Australian labour market is.

4.4 Some of Us Jumped on Our Scooters

One labour market trend thought to be hastened by the pandemic was the rise in the gig economy. While many of us pivoted to working from home (see below), many pivoted to a new kind of work. Those made redundant stood down or furloughed on reduced JobKeeper level wages looked to other opportunities. When stood down because of border closures and reduced domestic timetables, we read about airline pilots who had to find.

> many ways to fill their days and pay their bills while they wait out the pandemic. They've driven farm machinery, studied for law degrees, joined contact-tracing teams and stacked supermarket shelves (Cadzow 2021).

Others found opportunities in the gig economy, zooming around the streets of our cities on scooters or bicycles, delivering all kinds of things to those in lockdown. Although initially a casualty of the pandemic-induced economic slowdown, insecure self-employed positions became a lifeline of sorts for many. Research by The Australia Institute's Centre for Future Work (Nahum and Stanford 2020) shows that while 6% of workers classified as less-secure self-employed (including many working in the gig economy) lost employment during the initial stages of the pandemic, the sector saw significant growth in the months that followed.

Some evidence of the rise in gig workers can be gleaned from the Australian Bureau of Statistic's quarterly labour force data. Taking the August quarter of 2020 as an indication, the data tells us that while the number of employed people across Australia increased by 111,000 between July and August, a 0.9% increase, the number of people receiving payroll payments decreased by 1.2%. The commentary from the Bureau of Statistics tells us that non-employees accounted for a significant proportion of this growth. More specifically, they point out that.

> This growth in original employment can be attributed to an increase in the number of owner managers (self-employed people) without employees, most of whom work in an unincorporated enterprise (e.g. a sole trader), which increased by 50,200 (Australian Bureau of Statistics 2020b).

It may be true that some of the growth in self-employed workers might be accounted for by sub-contractors. But it is also highly likely that the change is accounted for by the increase in people forced into working in the gig economy, driving cars, delivering food, hanging up pictures and other odd jobs. Apparently, we can all be entrepreneurs now (Mitchell 2019).

The rise in non-employees becomes even more apparent when we consider the monthly changes in the employment index status of employment. Here the Australian Bureau of Statistics states.

> that changes in the number of employees and owner managers without employees were relatively similar between April and July. However, they diverged in August, with the number of owner managers without employees returning close to the level in March (Australian Bureau of Statistics 2020d).

Now while the companies running these gig platforms, as well as others, would argue that these gig workers are living some kind of dream, we all know that gig work is not what it is cracked up to be. Low pay, insecure contracts, no protection, no holiday allowances, and no sick pay. It sounds like a paradise, right?

A 2020 report of the Inquiry into the Victorian On-Demand Workforce (James 2020) referred to these people as 'low leveraged workers' who are.

> prominent in platform work: these include workers who are low skilled, more likely to be young and from a migrant background. They are operating in a labour market where there is high competition for fewer entry-level jobs (para 1314).

Furthermore, it suggested that.

> These are among the people most immediately impacted by the coronavirus (COVID-19) government interventions. Some are also more likely to be workers not eligible for the federal government's 'Job Keeper' payments, because they are irregular casual workers or visa workers (para 1315).

And pointed out that.

> These workers are in a precarious position in the labour market, especially at times of heightened unemployment or underemployment. Platforms offer them valuable opportunities to earn income, but with minimal choice about their arrangements, and low-income security (para 1316).

On top of the statistics about the growth in the gig economy, we also started to hear more disturbing facets of some forms of gig work. In September of 2020, we started hearing that, on the back of people whizzing around delivering take away pizzas, burgers or Chinese food on average, one food delivery rider had died in Australia every 11 days since 27 September. The media headlines read.

Australia's delivery deaths: the riders who never made it and the families left behind (Zhou 2020),

while the title of a podcast at the time read.

Dying to work: the dangers facing Australia's food delivery drivers (Murphy Oates and Zhou 2020).

One story (Zhou 2020) reported on a delivery rider who.

> died three days before the Melbourne lockdown lifted, holding someone else's food.
>
> [This person] had been in Australia for five years, having arrived from Malaysia …. First he was a student, then a chef, working part-time in a restaurant inside a casino. When the pandemic descended, like many other people around the world, he turned to food delivery—ferrying burgers and chips, burritos, and pizzas, across the city on a small motorised scooter. The car hit him on the corner of King and La Trobe streets at 7pm on a Saturday night.
>
> [He] was on a delivery when he died, which meant someone was expecting him. But for two and a half days, his death did not register in Australia (para 3-5).

To see people lose their lives while trying to eke out a living, earning wages considerably below the minimum without the usual workplace protections, is shameful. But, for many people left out in the cold because of being ineligible for government support, pivoting to these types of gig jobs may have been a necessary evil to help individuals and their families survive economically.

4.5 Some of Us Started Working in Our Pyjamas

While job loss was the new reality for many and had the potential to impact individuals and the communities they live in for some time, for others, the COVID-19 impact on their employment situation required nothing more than a shift of work location. The daily work commute became characterised by nothing more than going from the bedroom to the home office, kitchen table or sofa.

While working remotely was a trend that had begun to emerge in workplaces before the pandemic, initial moves towards working from home were slow to take off due to dominant management cultures that preferred a 'line of sight' (managers wanted to be able to keep an eye on us) and a view that face-to-face client and business dealings were superior for building a strong business culture (Holloway et al. 2020a). Before the arrival of COVID-19, around 8% of employees formally worked from home and worked from home for a median of one day per week (Productivity Commission 2021). With the arrival of the pandemic, normal workday routines were upended by the need for physical distancing to stop the spread of a highly infectious disease. Working from home is the rule rather than the exception for many.

Urban researchers such as Richard Florida and his colleagues (Florida et al. 2021) portray the lockdowns that became a common characteristic of cities as a representative of a forced experiment in work and commerce, while Australia's Productivity Commission, echoing similar sentiments argued that.

> The COVID-19 pandemic forced many workers and firms to experiment with working from home, showing that many jobs can be done well remotely (Productivity Commission 2021, p. 9).

The COVID-19 forced experiment was an exercise in testing teleworking, remote shopping and home deliveries that impacted differently on different individuals.

Across Australian cities, many jobs 'pivoted' to being carried out at home. The relatively little-known video conferencing technology 'Zoom' became as much a daily work tool as Word and Powerpoint. *Got a Zoom meeting* became part of everyday (virtual) office talk, and we began to read about zoom fatigue as online meetings started to take over people's calendars (Morris 2020). Working from home became so much of our daily existence that it even got its own hash-tag (#WFH).

The working from home trend was not just limited to cities and other large urban regions, but it was these places where it was most concentrated. Australians living in the Capital Cities were more likely to be working from home (36%) than those in country areas (25%). Social researchers at Roy Morgan (Roy Morgan 2020) estimated that during the first few months of the pandemic, 4.3 million people had been working from home. Later in the year, even as lockdown and stay at home orders eased, the Australian Bureau of Statistics reported that 31% of large business enterprises had more than 50% of their employees working from home, while small and medium-sized businesses had around one-fifth of their workforce at home (Fig. 4.8). Indeed, much of this book was written while working from our respective home offices.

For many, working from home had become the new normal, even after formal lockdowns had ceased and would continue after the pandemic had subsided (Hopkins and Bardoel 2020). A survey conducted by researchers at Swinburne University of Technology in May of 2020 reported that vast numbers of Australians anticipated they would continue to work at home in a post-pandemic world (Fig. 4.9). At least a quarter of those surveyed thought they would continue to work at home for at least two days a week, with another quarter saying they would work from home sometimes. It is clear that the working from home model won't be a 100% transition but will likely represent some hybrid or blended approach with.

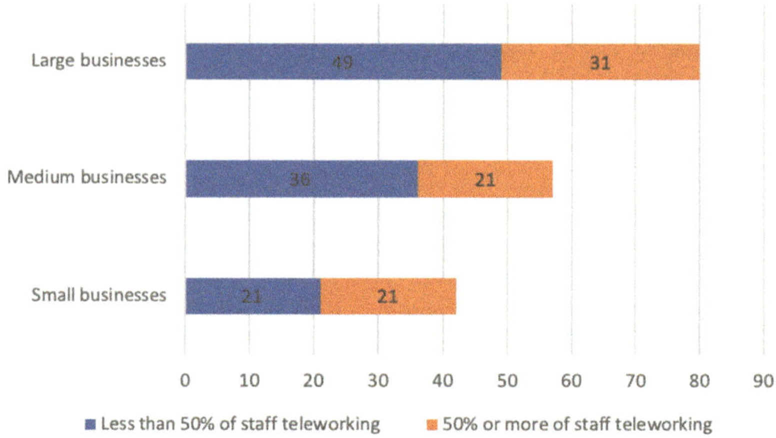

Fig. 4.8 Employing businesses with employees currently teleworking, by employment size (Australian Bureau of Statistics 2020a)

> allocate time slots—perhaps specific days—of in-office working for all employees to maintain workplace productivity and collaboration, while also allowing working from home to continue outside those hours (Nguyen 2021) (para 3).

But working from home wasn't for everyone. Not everyone could all of a sudden bring their work home. Those most likely to work at home were people employed in what many refer to as knowledge worker occupations or industries. These are people whose work comprises mostly abstract and/or cognitive tasks (Florida et al.

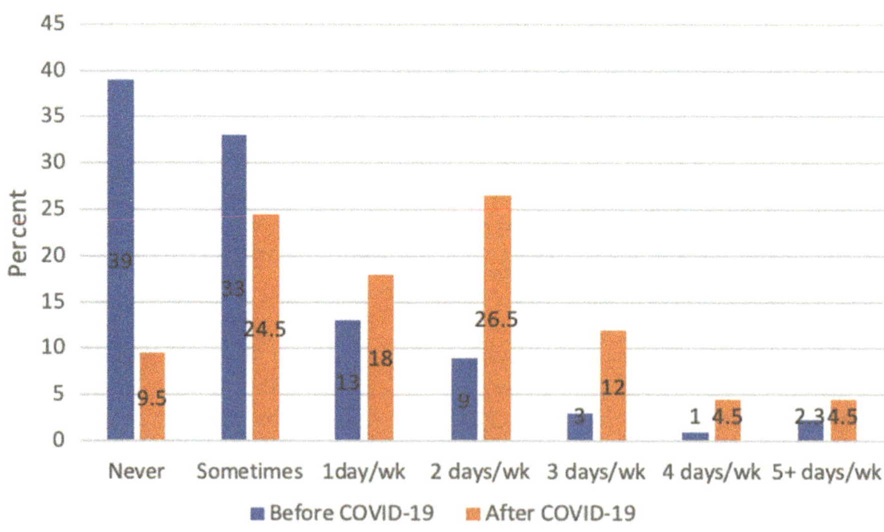

Fig. 4.9 Worker future work from home predictions (Hopkins and Bardoel 2020)

2021). Think, academics, lawyers, teachers, architects, accountants, design thinkers and others whose jobs involve thinking for a living. Others were also forced to work at home because of the forced experiment with differing degrees of success. The official picture from the Australian Bureau of Statistics was that workers in education and training, information media and telecommunications, financial and insurance services, professional, scientific and technical services and administrative and support services industries were more likely to have been working at home during the first six months of the pandemic (Fig. 4.10). These people could, more-or-less, work just as well in their lounge rooms as they could in their CBD offices or classrooms or where ever. It is less easy to bring your work home if you are a construction worker, a bus driver or a garbage collector.

Did people like working from home? They certainly liked the idea of a shorter commute; 89% of the Swinburne University of Technology claimed that a benefit of working from home was that they didn't have to face the daily commute. No more having to battle traffic, ride crowded busses or trains and no more parking hassles when you got to the other end. Many of those working from home also appreciated the greater flexibility WFH arrangements offered. Clock on when you want, have lunch when you want, wear tracksuit pants and slippers if you want. No water cooler chats about last night's episode of the latest reality television craze. If you were working from home, you could also easily zone out from boring meetings by turning off your camera and muting your microphone. Zoom would never tell your boss that you were

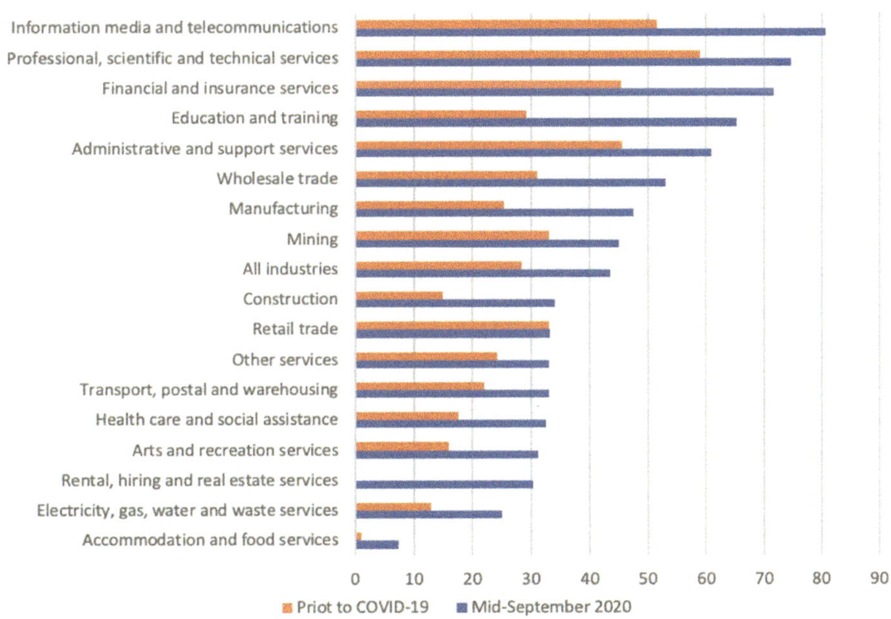

Fig. 4.10 Proportion of businesses with employees teleworking, prior to COVID-19 and mid-September 2020, by industry (Hopkins and Bardoel 2020)

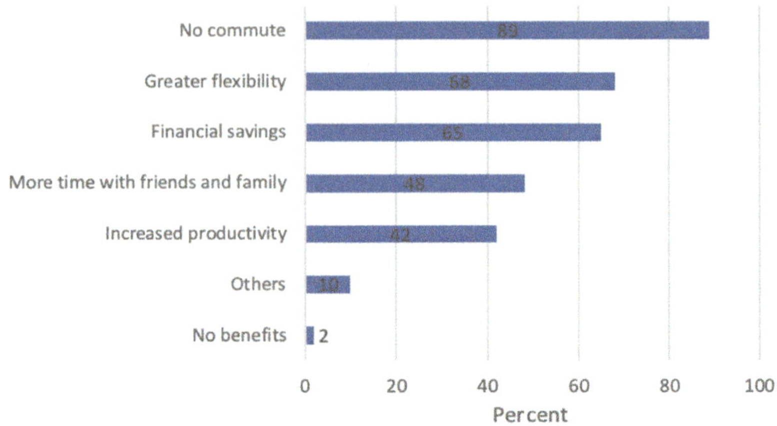

Fig. 4.11 Benefits of working from home during COVID-19 (Hopkins and Bardoel 2020)

actually doing a Sudoku puzzle rather than paying attention. The flexibility was, for many, endless. Other benefits included having more time with family and friends. The time spent commuting back and forth could now be spent playing backyard cricket with the kids or having a date night with one's spouse or significant other. The WFH trend also allowed people to save more, presumably due to lower transport costs and fewer café lunches, and despite what the early doomsayers predicted, working at home made workers more productive (Fig. 4.11). Even employers seemed to like it once they got used to the idea that working from home could actually produce productive outcomes.

But working from home wasn't all zoom meetings in your pyjamas and sleeping in till 8:59 am. For many, WFH wasn't all it promised. Work and home boundaries became increasingly blurred, and workers found it hard to switch off after they had 'left the office'. Then there was the myriad of potential distractions. That load of washing, that home improvement project, that new lockdown puppy or kitten. We have all been there. And if you were a parent with school-age children, the biggest distraction of them all was your sudden transition into a part-time teacher. Sure, we saved time getting the kids off to school in the mornings, but that saving was more than made up for by the constant need to remember high school maths or chemistry or find the latest bits and pieces for an art project that could not possibly wait. Never before had parents had a real appreciation for the work of teachers.

Some workers complained about a lack of motivation or a feeling of loneliness. A COVID-related feeling of alienation was a real thing for some, especially during long periods of working from home (this also impacted people who were on other stay at home orders). Although the exact extent of loneliness is yet to be comprehensively explored, psychologist Kevin Kniffen and his colleagues suggested that.

> The loss of social connections... is likely to negatively impact workers. Prior research has shown that high-quality social interactions—including informal chats among co-workers—are essential for mental and physical health. Handshakes that are also known to be valuable

> for social connection are now restricted. Against this backdrop, both the requirement to WFH and plans to de-densify workplaces in support of physical distancing are likely to have side effects that include at least some degree of harm to individuals' mental and physical health (Kniffin et al. 2021, p. 67)

For others, it was an infrastructure issue. Homes were turned into places with multiple work stations, pulled together at short notice (office supply stores made a killing) and often inadequately thought through with significant issues for any notion of home office OH&S. For those living in multi-person households trying to find adequate space for multiple home office setups was an issue. And despite all the hype surrounding the reliability and speed of Australia's National Broadband Network (NBN), unreliable Wi-Fi access, crucial for the whole working from home arrangement, was patchy for many (Fig. 4.12). Summing up the infrastructure issue, a report in the Australian Financial Review noted:

> Not everyone's experience of WFH is the same. A recent survey by software giant SAP ANZ, reported in the Australian Financial Review (Boddy 2020) found that while employee morale was positive overall, there were barriers to working from home for staff. One in four (26 per cent) said they did not have a computer desk, one quarter said they had problems connecting to internet networks at home, 23 per cent said they had limited private space, and one in five said they did not have an external monitor. Some employees do not have a dedicated space at home and feel embarrassed about the room they work in (para 18).

For Australian cities, the trend towards working from home has the potential to alter the traditional geography of jobs, at least in the short run. In their discussion of cities in a post-COVID world, Richard Florida and his colleagues (Florida et al. 2021, p. 11) pose the question.

> Can [working from home] create a new geography of labour, altering the fundamentals of economic geography and urban systems... This is probably the key aspect of the pandemic shock because it affects the overall economic structure and economic geography of work, concomitant residential choice behaviours, and the future of superstar metros and the urban system.

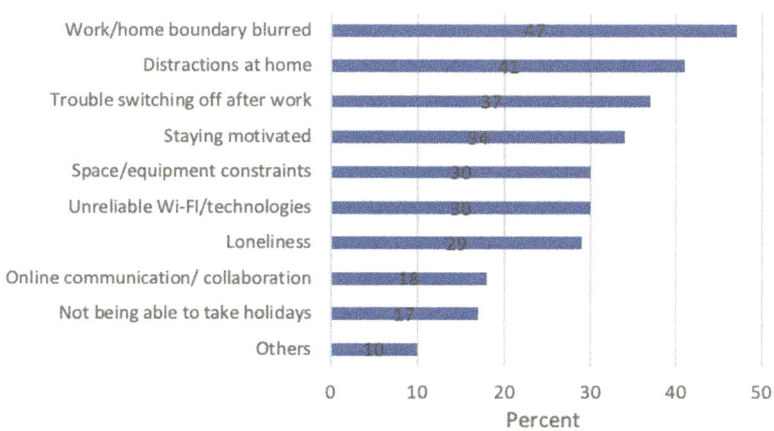

Fig. 4.12 Challenges of working from home during COVID-19 (Hopkins and Bardoel 2020)

Certainly, there is plenty of anecdotal and hard evidence to suggest that the future of work in Australian cities is changing and that the opportunities for a new working from home culture will vary across a diverse range of places and communities within our cities and towns. For Australian cities, researchers from SGS Economics and Planning (Holloway et al. 2020b) provide some analysis that goes to the heart of this question. Using data on the percentage of workers who regularly access the internet from home for their employers, the authors produced maps that allowed them to consider "those that are most likely to maintain the pattern post-COVID, and have an enduring impact on local places". Taking Sydney as an example (Fig. 4.13), the research identified clusters of locations, mostly in the eastern suburbs and the lower north shore, where significant numbers of workers regularly worked from home. This concentration was contrasted with locations in Sydney's outer-western suburbs and in the southwest, where much lower proportions of workers regularly worked from home. The other cities the SGS report (Fig. 4.14) considered had similar spatially distinct patterns of working from home activities leading the authors of the report to suggest that.

> If remote working continues at an increased rate, there is no doubt that it has the potential to fundamentally reshape the form and structure of our cities (para 11).

The potential impacts can be seen in the consequences for urban transport and have seen a rapid decline in traditional business hubs and CBD activity across Australia's large cities. As people pivoted to the home office, less and less travelled into major employment regions meaning that the support infrastructure that relied on a thriving

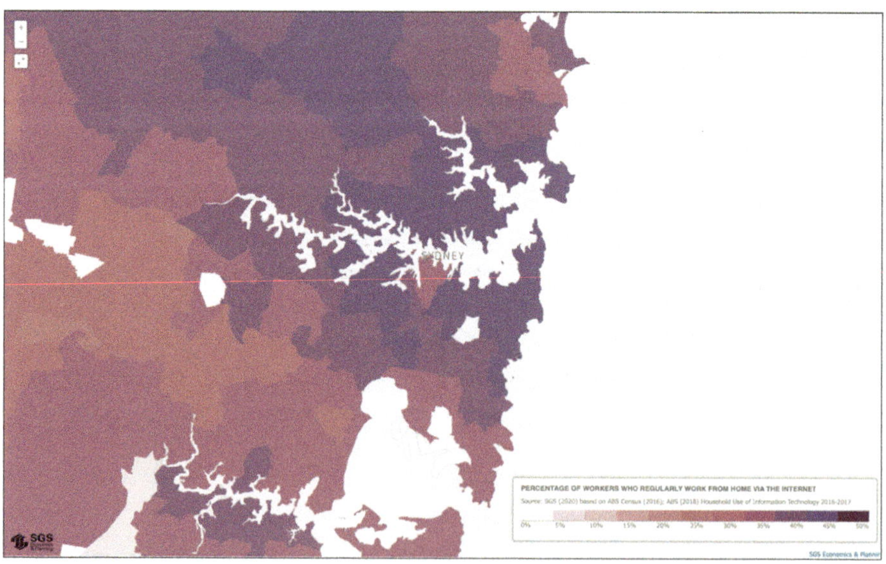

Fig. 4.13 Percentage of workers who regularly work from home via the Internet, Sydney (Holloway et al. 2020b)

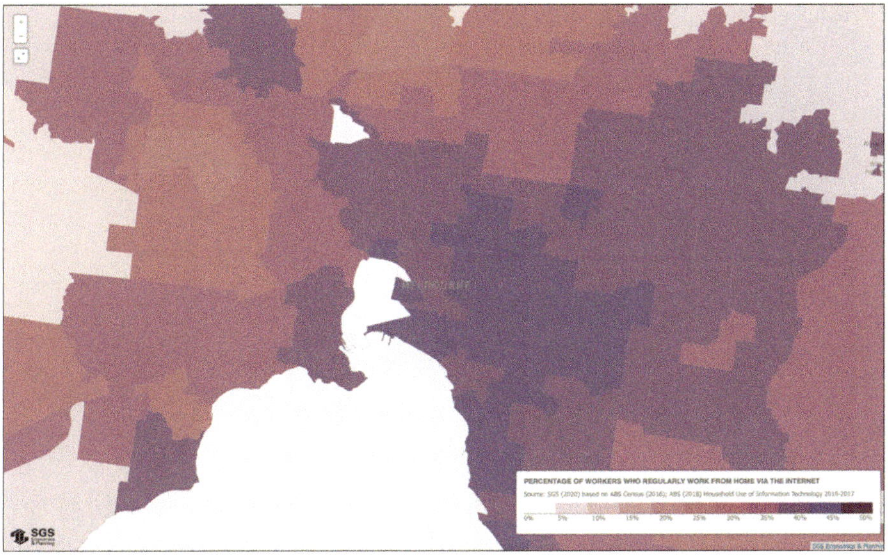

Fig. 4.14 Percentage of workers who regularly work from home via the Internet, Melbourne (Holloway et al. 2020b)

CBD took a hit. Anecdotally, we hear about catering companies who saw their commercial bookings disappear in the blink of an eye. And the ubiquitous inner-city coffee shop that sat beneath large office towers everywhere and was once booming now only saw a trickle of customers. Many of our CBDs and other concentrations of offices resembled ghost towns virtually overnight.

One particularly interesting piece of data to emerge during the COVID-19 pandemic was the mobility data released by Google. According to tech giant the

> Community Mobility Reports aim to provide insights into what has changed in response to policies aimed at combating COVID-19. The reports chart movement trends over time by geography, across different categories of places such as retail and recreation, groceries and pharmacies, parks, transit stations, workplaces, and residential (Google 2021) (para 2).

What the data illustrates is the changing level of activity in the central business districts of Australian cities. The Google data show percentage changes in visitor numbers from a baseline day: "the median value from the 5-week period Jan 3–Feb 6, 2020". So, we would expect to see a reduction in visitors to CBDs as the lockdowns and working from home came into force. Focusing on retail and recreation visits, researchers Paul Maginn and Gary Mortimer (Maginn and Mortimer 2020) considered this issue, and their results show clearly the rapid decline in CBD activity (Fig. 4.15). While all cities recorded increased visitors (compared to baseline) before the arrival of the pandemic, the data shows that.

> The arrival of COVID-19 in late February and government responses had a dramatic impact on visitors to retail/recreation places across all capital cities. CBD-dominant local government areas (LGAs)—Adelaide, Melbourne, Perth and Sydney—were more badly affected

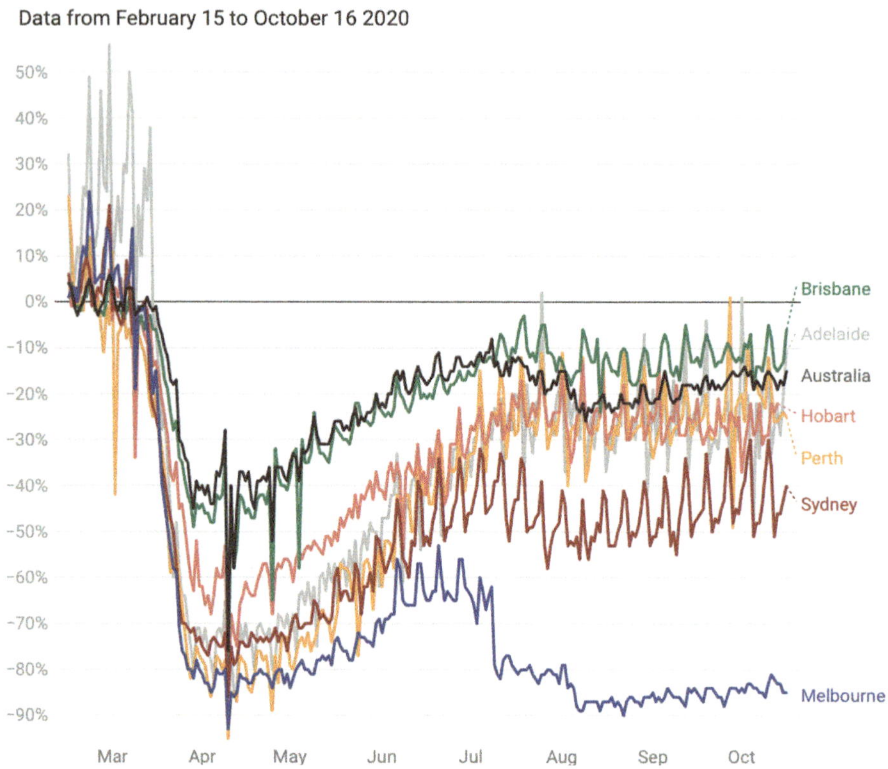

Fig. 4.15 Trends in retail/recreation visitor numbers in capital cities (Maginn and Mortimer 2020)

> than Hobart and Brisbane whose metropolitan regions are defined by a single LGA (Maginn and Mortimer 2020) (para 10).

From early to mid-March, the numbers went into free fall across all state capitals. The numbers in Perth crashed rapidly, down 42% on March 2nd. In the following week, Brisbane, Melbourne and Hobart fell between 10 and 34%. Sydney hit 19% on March 14.

This is bad, but we get a better picture when we look at the data that singles out CBDs and compares them with other localities (Fig. 4.16). By April 10 2020, the major CBDS had all witnessed falls, some at or close to 95%. Other localities fell too, but not nearly by the same magnitude. The Local Government Area of Sydney (the location of the CBD) witnessed a massive fall in the early stages of the pandemic lockdown and failed to make significant headway after that, thanks to strict provisions including stay at home orders and working from home mandates.

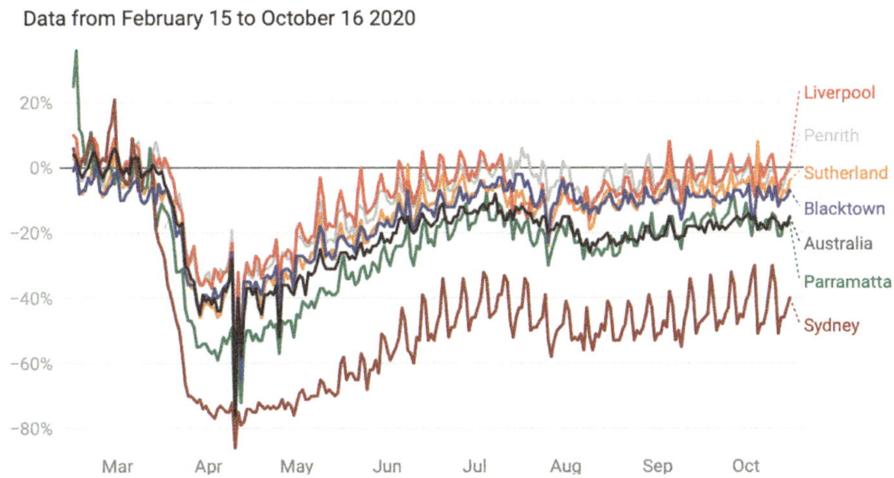

Fig. 4.16 Trends visitor numbers: Sydney CBD and outer suburbs (Maginn and Mortimer 2020)

4.6 Essential Workers: Not All Heroes Wear Capes

While there was a massive shift to working at home, it was not the case for everyone. For several occupations and the workers engaged in them, daily work life still required effort actually to make it to work. For these people working at home wasn't possible or desirable. Think about, for example, frontline health care workers, ancillary hospital staff (cleaners, security guards, etc.) and retail workers, or those employed in the transport, logistics, manufacturing and construction sectors. For many of these workers, their employment is location-specific or requires tools or specialised equipment that cannot be accessed from the home office.

During the early days of the pandemic, there was much debate about who was deemed an essential worker. Some were straightforward. Doctors, nurses and supermarket workers, for example. Others were more troublesome, and the politicians didn't help with the definition. New South Wales Health Minister Brad Hazard was quoted as suggesting that.

> To try and define essential work is really very challenging,
>
> An employer and their employee would know whether the worker is really essential (Davies 2021) (para 8).

Not very helpful. One report suggests that about 45% of workers can be defined as essential and are required to travel for work (Holloway et al. 2021). Some essential workers, including those working in infrastructure, construction and maintenance, might be considered non-public facing and can be contrasted with others in high-touch public-facing occupations that provide a range of services. These are the doctors, nurses, retail workers, cleaners and security guards that are all featured in the media as COVID-19 heroes, especially during the direst stages of the pandemic and associated lockdowns. Media reports trumpeted.

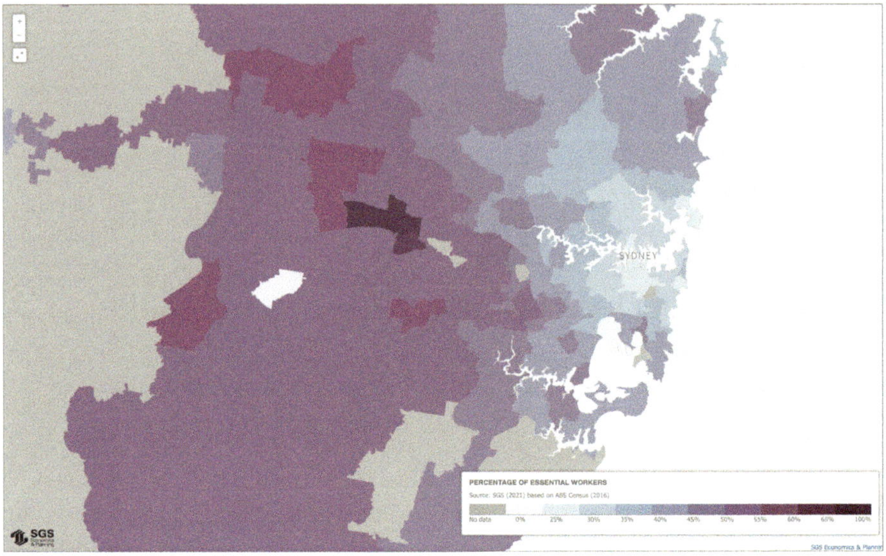

Fig. 4.17 Percentage of essential workers, Sydney (Holloway et al. 2021)

> Sydney frontline workers hailed as heroes amid Covid lockdown (Hudson-O'Farrell and Bharadwaj 2021).
>
> The quiet heroes who are keeping Canberra going during the pandemic (Evans 2021).

Because they must leave home to engage in work, essential workers and their families were exposed to a greater risk of contracting COVID-19, a fact often raised when nurses' aids, cleaners or security guards in quarantine facilities became infected. These particular workers' employment was often at multiple locations or with multiple employers, and they relied on public transport to travel between jobs. We noted in chapter three, that the COVID-19 infection hot spots in cities were often associated with low-paid workers who are unable to work from home. As we read in this SGS Economics and Planning report (Holloway et al. 2021), the pandemic has drawn attention to where our essential workers live. The report found that in Sydney, for example, it's mostly in the city's west and southwest (Fig. 4.17), while in Melbourne, the highest proportion of essential workers resides in the north and west growth areas on the city's outer edges (Fig. 4.18). These locations also recorded a high share of infections during this year's Delta outbreak.

4.7 What Now for the COVID Job Shuffle?

All of the labour market trends discussed in this chapter paint the picture of a highly disruptive event that, in many ways, spared very few people in terms of the ways they engaged or were disengaged from the world of work. Like the unequal burden

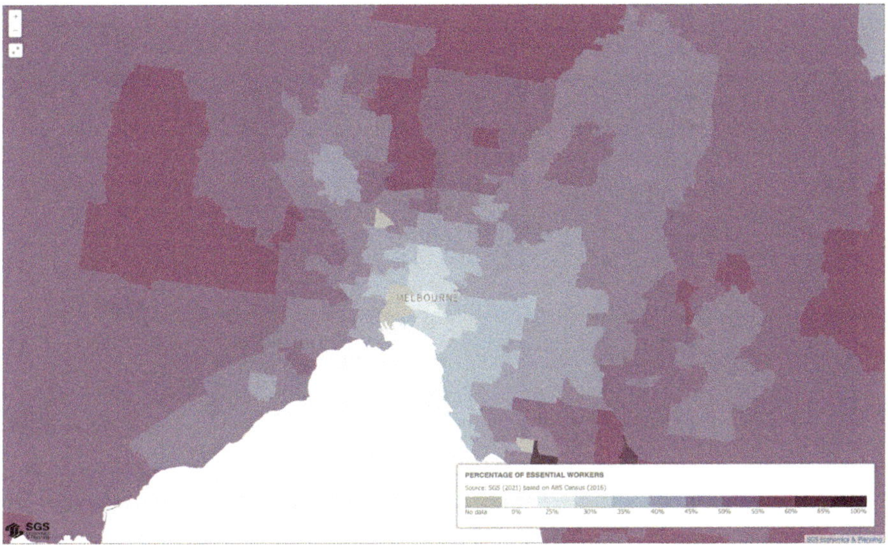

Fig. 4.18 Percentage of essential workers, Melbourne (Holloway et al. 2021)

we discussed in the previous chapter, the COVID-19 impact on jobs and how people worked was highly uneven. The longer-term consequences of the job's shuffle are still being played out. The national border shutdowns described in chapter two have meant that, for the short term at least, official unemployment rates have fallen, but there are also many people working in disadvantaged positions both in terms of job security and working conditions and in terms of being at the front line of any further disease outbreaks or variants. The job shuffle has changed how many people think and view their work life and has accelerated changes that have been bubbling away under the surface for some time. Only time will tell what the long-term consequences for our cities will be.

References

Anglicare (2021) Jobs availability snapshot 2021. https://www.anglicare.asn.au/wp-content/uploads/2021/10/Jobs-Availability-Snapshot-Report.pdf

Australian Bureau of Statistics (2020a) Business indicators, business impacts of COVID-19. https://www.abs.gov.au/statistics/economy/business-indicators/business-conditions-and-sentiments/sep-2020a

Australian Bureau of Statistics (2020b) Strong employment growth for non-employees. https://www.abs.gov.au/articles/strong-employment-growth-non-employees

Australian Bureau of Statistics (2020c) Weekly payroll jobs and wages in Australia. https://www.abs.gov.au/statistics/labour/earnings-and-work-hours/weekly-payroll-jobs-and-wages-australia/latest-release#state-and-territory

Australian Bureau of Statistics (2020d) Weekly payroll jobs and wages in Australia, week ending 22 august 2020d. https://www.abs.gov.au/statistics/labour/earnings-and-work-hours/weekly-payroll-jobs-and-wages-australia/week-ending-22-august-2020d

Australian Bureau of Statistics (2021) Labour force, Australia, Detailed. https://www.abs.gov.au/statistics/labour/employment-and-unemployment/labour-force-australia-detailed/latest-release

Baum S (2021) Using a regional lens reveals the uneven impact of the COVID employment crash. http://bilbo.economicoutlook.net/blog/?p=46861

Baum S, Mitchell W (2020) Employment vulnerability index 3.0. http://www.fullemployment.net/publications/reports/2020/EVI_3.0_Final_Report.pdf

Bishop J, Day I (2020) How many jobs did JobKeeper keep? Reserve Bank of Australia, RDP 2020, 7, 2020–2007

Boddy N (2020) Employee wellbeing could define post-pandemic leadership. Aust Fin Rev. https://www.afr.com/work-and-careers/careers/this-issue-could-define-post-pandemic-leadership-20200604-p54zbh

Butler B (2020) Coronavirus has delivered a hammer blow to Australia's economy. What happens next? The Guardian. https://www.theguardian.com/business/2020/mar/21/coronavirus-has-delivered-a-hammer-blow-to-australias-economy-what-happens-next

Cadzow J (2021) And the sunsets! If you're heading west, they go on forever': our pilots reminisce. Sydney Morning Herald. https://www.smh.com.au/national/one-day-we-looked-straight-down-into-the-eye-of-a-cyclone-20210809-p58h64.html

Chalmers S (2019) Lower interest rates, a slight growth pick-up, risks to the downside: where economists see Australia in 2020 [Press release]. https://www.abc.net.au/news/2019-12-23/where-economists-see-australia-in-2020-lower-rates-slow-growth/11796280

Creighton A (2020) Buckle up for hell of a bumpy ride: it's the response to COVID-19 that's generating an almost certain recession, rather than the virus itself. The Australian. https://www.theaustralian.com.au/inquirer/coronavirus-buckle-up-for-hell-ride/news-story/5101764f3578c685a69e819977280207

Davies A (2021) Words fail Berejiklian government as 'essential' NSW workers remain a mystery. The Guardian. https://www.theguardian.com/australia-news/2021/jul/13/words-fail-berejiklian-government-as-essential-workers-remain-a-mystery

Evans S (2021) The quiet heroes who are keeping Canberra going during the pandemic. The Canberra Times. https://www.canberratimes.com.au/story/7404888/the-quiet-heroes-who-keep-canberra-going/

Florida R, Rodríguez-Pose A, Storper M (2021) Cities in a post-COVID world. Urban Stud 00420980211018072

Google (2021) COVID-19 community mobility reports. https://www.google.com/covid19/mobility/

Holloway A, Mackevicius L, Zierke M, Allen T, Ratnam K (2020a) Where are people working from home, and how could this reshape Australia's cities and regions? https://www.sgsep.com.au/publications/insights/where-are-people-working-from-home-and-how-reshape-cities-and-regions

Holloway A, Mackevicius L, Zierke M, Allen T, Ratnam K (2020b) Where are people working from home, and how could this reshape Australia's cities and regions? https://www.sgsep.com.au/publications/insights/where-are-people-working-from-home-and-how-reshape-cities-and-regions

Holloway A, Szafraniec J, Hoang M, Keegan M (2021) Closing the divide: essential workers, Australian cities and COVID-19. https://www.sgsep.com.au/publications/insights/closing-the-divide-essential-workers-australian-cities-and-covid-19?utm_medium=email&_hsmi=178997769&_hsenc=p2ANqtz-_cM0kz0Au0ZweLKSJtPO4vkr7EuHUtmfRlPj2cLw_plfe2m_cx1lJiLZqjfcE7qD-JsolwtXtfadbmfU8X3Qf3oNa82h7UgZugLs2cnwV3oo3jt0E&utm_content=178997769&utm_source=hs_email

Hopkins J, Bardoel A (2020) Key working from home trends emerging from COVID-19

Hudson-O'Farrell K, Bharadwaj A (2021) Sydney frontline workers hailed as heroes amid Covid lockdown. The Daily Telegraph. https://www.dailytelegraph.com.au/coronavirus/sydney-frontline-workers-hailed-as-heroes-amid-covid-lockdown/news-story/1c33d83946425a5247c2be86b2080faa

James N (2020) Report of the inquiry into the Victorian on-demand workforce. Victorian Government, Melbourne

Janda M (2020) Australia in its first recession in 29 years as March quarter GDP shrinks

Kniffin KM, Narayanan J, Anseel F, Antonakis J, Ashford SP, Bakker AB, Choi VK et al (2021) COVID-19 and the workplace: implications, issues, and insights for future research and action. Am Psychol 76(1):63

Maginn P, Mortimer G (2020) How COVID all but killed the Australian CBD. The Conversation. https://theconversation.com/how-covid-all-but-killed-the-australian-cbd-147848

Mitchell W (2019) We are all entrepreneurs now marching towards a precarious and impoverished future. http://bilbo.economicoutlook.net/blog/?p=42447

Mitchell W (2020) Australia labour market—getting worse as unemployment and underemployment rise. http://bilbo.economicoutlook.net/blog/?p=44343

Mitchell W (2021) Australian labour market goes backwards in April. http://bilbo.economicoutlook.net/blog/?p=47509

Morris B (2020) Why does Zoom exhaust you? Science has an answer. Wall Street J 27

Murphy K (2020a) Australia enters first recession in 29 years after blow from bushfires and coronavirus. The Guardian. https://www.theguardian.com/business/2020a/jun/03/australia-enters-first-recession-in-29-years-after-blow-from-bushfires-and-coronavirus

Murphy K (2020b) Australian federal treasurer Josh Frydenberg presents the nation's latest GDP data showing the economy is in recession. The Guardian. https://www.theguardian.com/business/2020b/jun/03/australia-enters-first-recession-in-29-years-after-blow-from-bushfires-and-coronavirus

Murphy-Oates L, Zhou N (2020) Dying to work: the dangers facing Australia's food delivery drivers, The guardian, https://www.theguardian.com/australia-news/audio/2020/nov/13/dying-to-work-the-dangers-facing-australias-food-delivery-drivers

Nahum D, Stanford J (2020) 2020 year-end labour market review: insecure work and the COVID-19 pandemic

Nguyen M (2021) Research shows working from home doesn't work. Here's how employers should tackle the problem. Time. https://time.com/6088110/remote-work-structured-hybrid-research/

OECD (2020) Coronavirus: the world economy at risk

Productivity Commission (2021) Working from home: research paper. https://www.pc.gov.au/research/completed/working-from-home/working-from-home.pdf

Roy Morgan (2020) Nearly a third of Australian workers have been '#WFH' [Press release]. http://www.roymorgan.com/findings/8451-roy-morgan-working-from-home-june-2020-202006290638

Sandbu M (2020) The economics of belonging. Princeton University Press

Select Committee on COVID-19 (2020) Select Committee on COVID-19 first interim report. Parliament of the Commonwealth of Australia, Canberra, Australian Capital

Uren D (2020) The unequal Burden of the COVID-19 labour market collapse. https://www.ussc.edu.au/analysis/the-unequal-burden-of-the-covid-19-labour-market-collapse

Zhou N (2020) Australia's delivery deaths: the riders who never made it and the families left behind. The Guardian. https://www.theguardian.com/business/2020/nov/21/australias-delivery-deaths-the-riders-who-never-made-it-and-the-families-left-behind

Chapter 5
Housing and the Pandemic

5.1 Housing Outcomes in the Pandemic City

Cities and housing go hand-in-hand. Housing consumes a significant amount of space within cities and adds significantly to the urban feel and character. Adequate housing remains closely connected with economic development. It generates employment, promotes industries and creates opportunities for financial gain. It can also make the perfect storm if households find themselves in a falling market with increasing financial burdens. When the economy unravels, people's housing dreams can disappear. Even in regular times, housing, particularly secure housing, plays a vital role in the well-being of individuals and families in society. It protects from the elements and is where significant social interaction occurs. It is

> critical to our survival. It plays an integral role in promoting or undermining not just our health and safety but also our mental well-being. Housing is also intrinsic to our sense of self-esteem and our perceived control over our surroundings and has the potential to directly influence a range of other outcomes, from education to employment to social participation. Such is the centrality of housing to each person's day-to-day life that it invariably plays an important role in shaping how we see our own lives and our place in the world around us (Coates et al. 2015, p. 2).

All of this makes housing closely intertwined in our daily lives. The arrival of the COVID-19 pandemic made it even more so. The sudden need to lockdown, to be able to social distance and to self-isolate meant that residential accommodation became much more than just a place to spend time outside of activities such as work or school. It became for most people their whole world: their workplace, school, playground and place to socialise, albeit at an appropriate distance. We saw the home offices become the new norm (Chap. 4), and we saw an increase in parents juggling their working life with home-schooling activities. People set up home gyms and neighbourhood parties with everyone staying in their respective driveways, becoming the norm. Even traditional ANZAC day dawn services, the day that some say forged the Australian spirit, were relegated to the safety of one's own home rather than the usual mass

S. Baum et al., *Pandemic Cities*, Cities Research Series,
https://doi.org/10.1007/978-981-19-5884-7_5

gatherings at significant places of remembrance. The old saying, *be it ever so humble; there's no place like home*, had never been so true.

The onset of COVID and its accompanying economic slowdown had many commentators and industry players concerned about the potential for a housing market collapse. But thanks to government interventions, including the federal government's HomeBuilder scheme and First Home Loan Deposit Scheme, together with a raft of state-based grants and stamp duty concessions for first home buyers and record-low interest rates, the aggregate housing market appeared to flourish. House prices rocketed around 25%, following an initial dip, while rentals likewise saw increases by almost 12% (good news for investors, not so good for renters) (Owen 2022). During 2020, despite the angst around a fall in building activity, new building commencement of residential buildings saw an increase each quarter, only slightly dropping off towards the end of 2021 (Australian Bureau of Statistics 2022). You would be wrong to think that we dodged a housing bullet during the pandemic despite all of this. No. While there were some winners thanks to the booming market, many people whose housing situation was associated with stress, COVID risk and generally poorer outcomes. Like many of the discussion points in this book, the pandemic revealed the inequalities and disadvantages in our cities.

5.2 COVID-19 Adding to the Housing Crisis

For Australia, even before the onset of COVID-19, the housing situation was one of marked inequalities and crises. While many individuals and households could access housing that was secure, affordable and suitable to their needs, many others faced hardship in their housing choices, which severely limited choice or forced them out of secure housing altogether. The arrival of the pandemic impacted Australia's housing situation, exposing its many flaws in much more stark ways than what was previously the case.

The most pressing housing questions associated with COVID-19 in Australian cities revolve around housing's intersection with human health and well-being and with the role of housing as a commodity with varying financial implications for households at different stages of the property ladder. Without access to suitable housing, health and well-being are compromised. We know this from the epidemiological data and the boundless studies into the risks posed by inadequate housing (Bonnefoy 2007). As with other things, COVID-19 magnified these issues and led to multiple concerns for at-risk populations. COVID-19 also interfered with the financial equations associated with housing access. The negative economic impacts (discussed in Chap. 4) resulted in an extension of earlier housing affordability issues, which either put housing out of reach or placed households into perilous positions, living in unsuitable accommodation as housing prices and rents increased while average Australian real wages declined.

Chapter 3 outlines the links between COVID-19 and the social and economic inequality that characterise our cities and other large urban areas, arguing that the pandemic had made Australia's social inequalities even starker. A vital piece of that puzzle is the housing system, whose ability to divide households into winners and losers only increased during the pandemic. As many researchers have pointed out, unaffordable, crowded and marginal housing (including homelessness) reflects some of the fundamental drivers that, when added to individual risk factors, multiply the potential for disease transmission and spread (Hu et al. 2021; Tinson and Clair 2020).

5.3 Marginal Housing and Homelessness During COVID-19

The arrival of COVID-19 in our cities sounded an additional layer of caution for those caught in homelessness and marginal housing and those at the sharp end of trying to help them. We all know what homelessness is. It's the person sleeping rough in the park or under the bridge. Or it's the family living out of their car. But it also includes persons staying temporarily with other households or people living in boarding houses or other temporary lodgings, or even people living in severely crowded dwellings. The Australian Bureau of Statistics' (2012) definition of homelessness, which they use for statistical analysis, is

> … when a person does not have suitable accommodation alternatives, they are considered homeless if their current living arrangement:
>
> - is in a dwelling that is inadequate;
> - has no tenure, or if their initial tenure is short and not extendable; or
> - does not allow them to have control of, and access to space for social relations'(para 2).

The issue of homelessness in Australian cities has received varying attention in the public eye. But, out of all populations trying to survive in our pandemic cities, those identified as homeless or residing in some other forms of marginal housing are the ones that were often hit the hardest. Public Health measures, including social distancing, restricting movements and gatherings and stay-at-home orders, have disproportionately impacted those without secure housing. Once public health orders were enacted across Australia, people living in many forms of marginal housing were faced with the paradoxical demand of staying at home when often they literally had no home in the traditional sense or, at best, had a house that made it so difficult to comply with preventive measures that it was next to useless (Crouzet et al. 2022).

The problem with understanding the homelessness situation is that it is not always visible. While the extreme of sleeping rough can often be seen in our city's inner-city parks or other areas, crowded housing is harder to identify to the casual observer, as are other forms of insecure housing such as short-term boarding accommodation. But we know that there is a sizable population of people living in these situations.

Estimates by the Australian Bureau of Statistics placed the number of homeless people at 49.8 people per 10,000 population. In absolute numbers, the 2016 census

Table 5.1 Number of homeless persons, by homelessness type, Census 2016. *Source* ABS 2018 (Australian Bureau of Statistics 2018)

Type of homelessness	Number
Persons living in improvised dwellings, tents, or sleeping out (rough sleepers)	8,200
Persons in supported accommodation for the homeless	21,235
Persons staying temporarily with other households	17,725
Persons living in boarding houses	17,503
Persons in other temporary lodgings	678
Persons living in severely crowded dwellings	51,088
All homeless persons	*116,427*

estimated that more than 116,000 people were homeless across Australia when the count was taken. Over half were male, 21% were aged 25–34 and 20% identified as Aboriginal and Torres Strait Islander Australians (Australian Bureau of Statistics 2018). Around 51,000 or 44% were living in severely crowded dwellings, with another 21,000 or so (18%) living in supported accommodation for the homeless. Over 8,000 were rough sleepers. (Table 5.1).

Early in the pandemic, the news cycle picked up on the struggle homeless people faced as the disease took hold in the community.

> The COVID-19 pandemic has made life on the streets more difficult than ever.
>
> All the facilities have closed down, so we've got nothing to do…You hear many people are dying from this virus, and I've had chest infections, I've had pneumonia, I'm very vulnerable to this disease. If I get it, I'm buggered. I could die from it.
>
> It's scary, but what can I do? (Baker 2020) (para 5–6).

The effective shutdown of cities meant that not only did the central business districts close down, but councils closed public facilities such as libraries, public toilets and public use barbecues which people sleeping rough often rely on to survive day-to-day.

The characteristics of homelessness and other forms of marginal housing make it especially problematic in a pandemic such as COVID-19. Severely crowded housing is a serious problem in terms of pandemic control. Before the pandemic, the World Health Organisation showed that crowding could significantly impact physical and mental health. In terms of infectious diseases such as COVID, the inability to effectively isolate increased the risk of spread (World Health Organisation 2018). With the release of COVID-19 case data, we have seen across a host of research (Hashim et al. 2020; Varshney et al. 2022) the strong associations between overcrowding and the risk of COVID. In our analysis (see Chap. 3), living in crowded dwellings was associated with higher rates of COVID infections and higher rates of mortality. For those living in highly overcrowded homes, the risks were high. The very nature of overcrowded housing meant it was tough to social distance, a critical preventative action to stop the spread. If someone caught the disease, the closeness of living arrangements meant that it could quickly spread to others in the same house. But it is not just crowding that is at issue. People may not be living in crowded housing arrangements, but it may be the standard of the housing they live in that poses a risk. In shared housing or boarding houses, residents are forced to share facilities, including kitchens and bathrooms, making it challenging to self-isolate and increasing the risk of contracting the disease when other residents have been exposed.

As day-to-day living became more challenging for the homeless and those in other forms of marginal housing, the COVID risk increased. In Chap. 3, we saw that the dangers associated with COVID were not evenly spread across the population. As we noted there,

> On various occasions, the phrase 'COVID-19 does not discriminate' has been repeated. This, however, is a dangerous myth, sidelining the increased vulnerability of those most socially and economically deprived (Patel et al. 2020, p. 110).

A lot of the discussion around the disease has focussed on the vulnerability of older populations. At the beginning of the pandemic especially, it appeared to be older people who were catching COVID, and it was older populations who had a higher mortality risk if they did catch it (Kang and Jung 2020). If you were an older person living in a marginal housing situation or homeless, your risk multiplied. Older people who are homeless or in marginal housing are more likely than others to experience poor physical and mental health. They can also experience earlier onset of health issues often associated with advanced age (Australian Institute of Health and Welfare 2021b). The risk of catching COVID-19 and the mortality implications represent a serious and concerning issue for these people.

When the five-year census rolled around in 2016, the Australian Bureau of Statistics estimated that 18,600 people aged 55 and over were homeless, accounting for almost one-fifth of Australia's homeless population. A further 13,600 older Australians were living in marginal housing, including caravan parks and crowded or improvised dwellings, who were considered at risk of becoming homeless. The most common form of homelessness for older people involved living in boarding houses (27%) or staying temporarily with other households (24%) (Fig. 5.1).

A critical sticking point in public health practices surrounding the pandemic was that, as we saw in Chap. 4, the option to work from home was not viable for many people. For homeless people and those in crowded shared accommodation, this raised issues. Not being able to find suitable space at home to work or home school was mentioned as a disadvantage by some people regardless of the type of accommodation they lived in (see Fig. 4.12, Chap. 4). Imagine what it is like if your home happens to be your car or a tent under a bridge or if you share a packed house with a group of unrelated people. Besides this issue, there was the opposite problem whereby those employed in the so-called essential services occupations (health care workers, social services workers, retail workers, etc.) didn't even have the option to work from home. The challenge posed by homelessness and marginal housing, in this case, was different. Those who worked in jobs that placed them in close proximity to others potentially faced greater exposure risk to the virus. Then their housing situations multiplied the problem by potentially spreading the virus to others in the household (in the case of shared accommodation or boarding houses) due to the

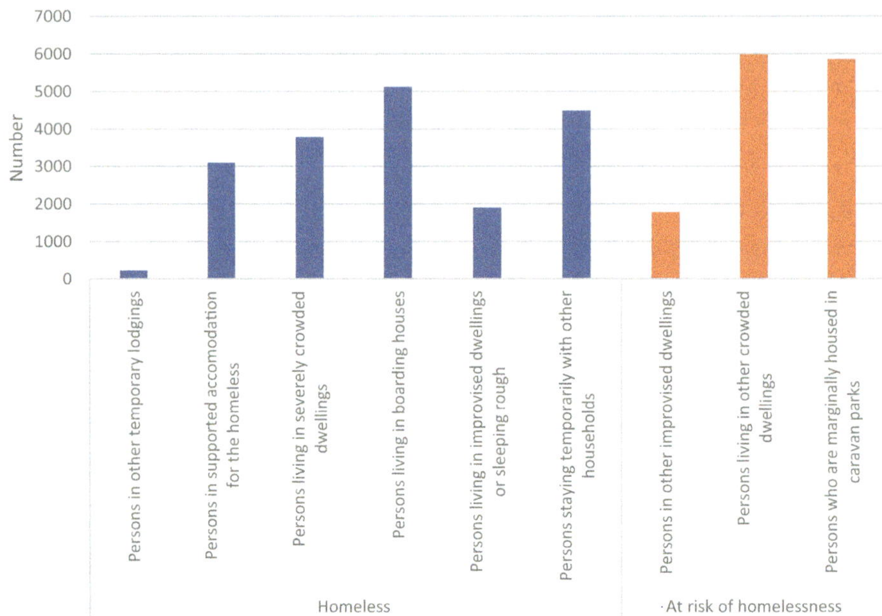

Fig. 5.1 Older people experiencing homelessness or at risk of homelessness by type of dwelling, 2016 (Australian Institute of Health and Welfare 2021b)

difficulty surrounding self-isolation or the inability to access appropriate help in the case of those sleeping rough.

An Australian Housing and Urban Research Institute report released in 2020 (Buckle et al. 2020) places some context around this issue. The report noted that in Sydney, at the time of the 2016 Census, 963 healthcare and social assistance workers lived in what might be referred to as severely crowded housing situations. Additionally, another 8,746 of the same workers lived in group households, and 3,592 were renting in multi-family households. The figures for Sydney were likely played out in cities across the country with potentially serious outcomes.

5.3.1 *Governments to the Rescue*

But the coronavirus shutdown wasn't all bad news for those in the most disadvantaged housing situations. While many people fell through the cracks, there was a focussed effort across Australia's largest cities to move people sleeping rough off the streets and into rooms in hotels. State governments began utilising hotels left empty following the effective shutdown of tourist activities as accommodation for those without secure permanent housing. Research by the Australian Council of Social Services (Pawson et al. 2021) estimated that by September 2020, at least 40,000 people had benefited from this emergency housing provision. During this time, the governments of New South Wales, South Australia and Victoria enabled funding to meet the costs of emergency accommodation, including hotel room charges, meals and the provision of social support in the form of support workers employed through non-government organisations (Pawson et al. 2021). While the provision of emergency accommodation could have easily been treated as a temporary stop-gap measure, governments also made attempts to introduce more sustainable solutions, including providing support for transition into more stable long-term housing solutions.

5.4 COVID-19 and Rental Financial Stress

While living in marginal housing introduces an extra layer of COVID risk factors, the arrival of the pandemic also added a financial problem as it threatened to worsen the already spiralling rental affordability crisis. As with a range of other social and economic issues, the arrival of COVID-19 and the public health and related financial challenges threw the problems of Australia's rental market into sharp relief. We have already discussed the impacts of the COVID-induced economic slowdown on the ability of some groups to maintain financial security. Social surveys, especially in the early days, told of the lengths some individuals and households had to go to due to an actual or potential drop in income. For example, a Survey by the Melbourne Institute on the impacts of COVID-19 found that in the early weeks of the pandemic, around 30% of the survey respondents felt financially stressed, with the highest numbers being in the ages of 18–44 years with approximately two in five (44%) experiencing moderate to high levels of financial stress (Melbourne Institute: Applied Economic & Social Research 2020).

An Australian Housing and Urban Research Institute report released in 2020 clearly illustrated the dire situations some renters found themselves in. Using data collected in the middle of 2020, the researchers paint a picture that suggests that financial stress was a significant issue for many in the private rental sector and had worsened during the initial stages of the pandemic. Many people they spoke to had seen a change in their employment situation or income due to the pandemic. Almost 1 in 5 had either their hours or income reduced, while 10% reported they had experienced a complete loss of income and 13% had suffered temporary job loss (Baker et al. 2020). These changes left many renters in a precarious financial position. The researchers found that just over 40% of the people they spoke to had experienced some form of negative change to their living situation due to COVID, with a large (22%) number reporting more than one change. Struggling to make ends meet was the big-ticket item (40% mentioned this), with just over a quarter having to skip meals to save money. For others, rent simply became unaffordable (15%), while some had to find a housemate to help cover the rent or moved out altogether (Fig. 5.2).

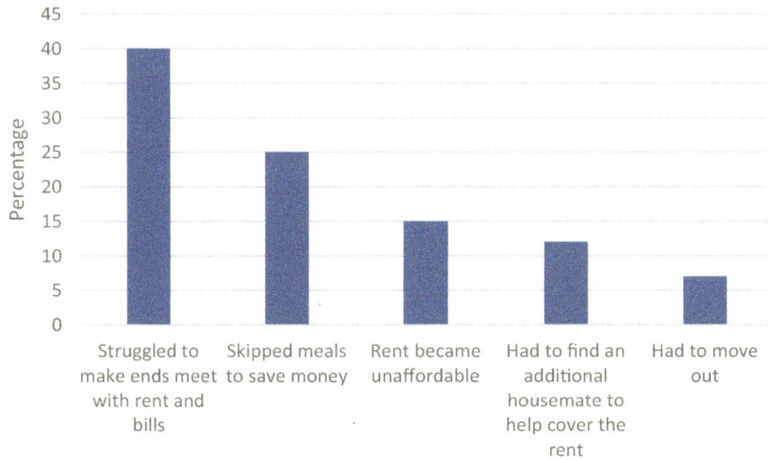

Fig. 5.2 Changes in the living situation of respondents (Baker et al. 2020)

These challenges were also reflected in the comments made to the researchers by the people they spoke to. When asked "as a result of COVID-19, has your living situation been affected" typical comments included

> Have sold most of my personal belongings to put food on the table for my family
>
> Moved back home to my Mum's
>
> Skipped meal to make sure the children were fed
>
> Charities for food hampers, and vouchers so I could afford bills
>
> Having to ask parents for help with money (Baker and Daniel 2020)

The potential impacts of increasing financial difficulty during the early stages of the pandemic were likely, as we have seen to be significant across several areas. But on the housing front and especially in the private rental market, any drop in earnings could place renters under financial stress. Generally referred to as the 30/40 rule, those in housing financial stress are lower-income households (lowest 40% of income) that spend more than 30% of gross household income on housing costs. For those in the rental market, estimates vary, but even before the pandemic, it was thought that about 43.1% of low-income households faced rental financial stress. The figure was higher in the capital cities than elsewhere, sitting close to 50% (Fig. 5.3) (Australian Institute of Health and Welfare 2021a).

The pandemic and its associated economic hit introduced the potential for these numbers to magnify. And magnify significantly. There was a host of anecdotal evidence that renters were doing it tough. Newspaper reports during the pandemic stated,

> Rental stress on the rise as low-income Australians priced out of the market (Convery 2021)
>
> Victorian renters struggle to afford food and risk eviction; a year into the COVID-19 crisis (Redman 2021)

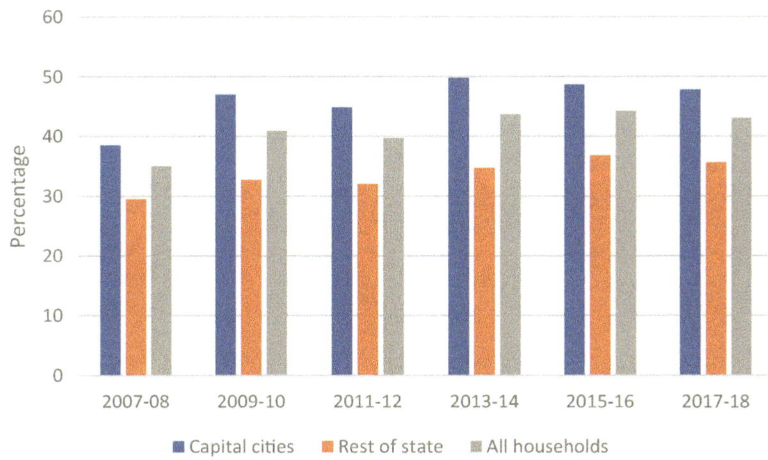

Fig. 5.3 The proportion of low-income households in rental stress, by location, 2007/2008 to 2017/2018 (Australian Institute of Health and Welfare 2021a)

It is not surprising that some households were struggling in the rental market. A report by consultants SGS Economics and Planning (SGS Economics and Planning 2020) provided a snapshot of rental financial stress during COVID. The curious thing about rental affordability during the pandemic was that, in aggregate, Australia's capital cities witnessed an improvement in rental affordability. But as with all things, aggregates hide a lot of subtle differences that are exposed once you start unpacking the data. One of the questions the researchers at SGS pondered was who was facing problems regarding affording rents across our cities? Often it was the usual suspects—single people trying to live on JobSeeker payments, single pensioners, part-time workers and those on minimum wages. Given that the COVID-19 economic impact had increased the financial precariousness of many of these groups, the effect on their rental financial stress was predictable.

Not surprising entire swaths of our cities were largely out-of-bounds to low-income renters. Take single people on JobSeeker allowance. Although the federal government temporarily increased the level of income support this group received via the so-called 'Coronavirus Supplement', it was still the case that rents across the country were out of reach for many. The SGS data found that young singles in this group would need to pay between 42 and 69% of their income to afford the median rental in any of Australia's capitals. For example, maps of our two largest cities showed significant areas that were out-of-bounds (Fig. 5.4).

Although they benefited from a temporary increase in income support, single pensioners were another group who still faced significant rental affordability issues. The SGS data found that across the capitals, a significant 50% of a pensioner's income would be needed to rent even a one-bedroom dwelling, introducing extra cost-of-living pressures on top of other costs such as the health care costs associated with ageing. The pressure was even greater for those trying to live in the Sydney rental

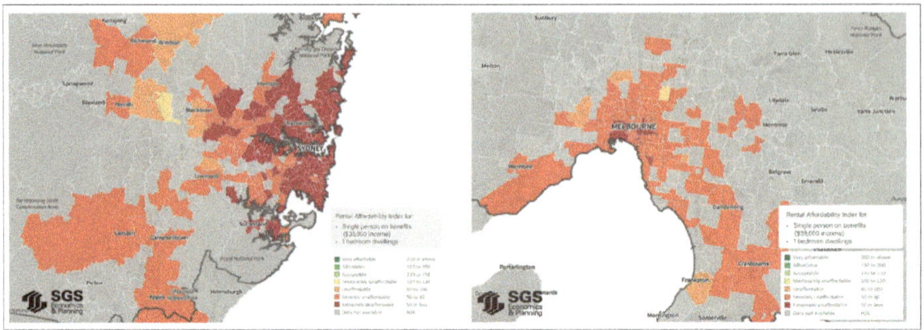

Fig. 5.4 Rental affordability index, single person of jobseeker payments, Sydney and Melbourne, June Quarter 2020 (SGS Economics and Planning 2020)

housing market. An estimate of 79% of income was needed, making significant city areas out of reach (Fig. 5.5).

Even if you were lucky enough to be working, if you were only receiving minimum wages or were only employed part-time, your rental financial risk still posed an issue. One particular group at risk was single parents. The SGS report identified that although this group was eligible to receive the extra coronavirus supplement, rents across the metropolitan areas were largely unaffordable. As with other household types, this group also faced increasing cost-of-living pressures associated with the need to provide for their families, which worsened the situation. In Greater Sydney, for example, single parents working part-time and receiving benefits would have to spend nearly half their incomes to gain access to a median-priced rental property, making large parts of the metropolitan region inaccessible (Fig. 5.6).

As with many of the other issues dealt with in this book, these affordability issues or the patterns they portray are not new. What the COVID-19 pandemic has done in many cases is to multiply the disadvantage and, in many cases, make it much more visible. The impact of increased financial stress also added another risk factor in the

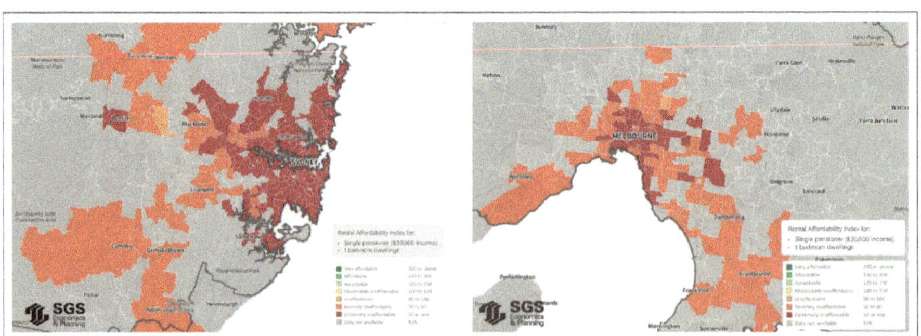

Fig. 5.5 Rental affordability index, single pensioners, Sydney and Melbourne, June Quarter 2020 (SGS Economics and Planning 2020)

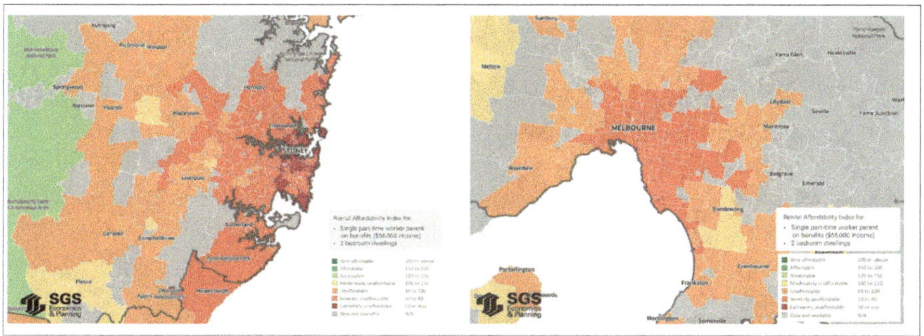

Fig. 5.6 Rental affordability index, single parent working part-time on benefits, Sydney and Melbourne, June Quarter 2020 (SGS Economics and Planning 2020)

time of COVID. One of the critical public health measures associated with COVID was that you should stay home and self-isolate if you were sick. That was the health advice from the very early stages of the pandemic. But, a sticking point in the plan was that for many individuals, and likely those facing rental financial stress, a lack of any financial buffer meant that even if they were sick, they still needed to attend work to make ends meet. This then acted as a potential increased risk of community transition as they went about their daily work routine.

Like groups caught up in marginal housing situations and homelessness, many private renters, especially those suffering the financial impact of rising rents and falling income, face a complex double whammy when it comes to COVID. To keep their head above water, some struggling private renters may have resisted public health orders to self-isolate if they became ill and continued working or if they worked in an essential occupation and had to put their health at risk by continuing to work to keep a roof over their heads. Combined with this, we know that the mental anguish caused by the threats posed by financial problems is likely to be especially pertinent for private renters caught in financial difficulty. The Australian Housing and Urban Research Institute report identified that several housing-related issues impacted renters during COVID. Emma Baker and her colleagues (Baker et al. 2020) found that around 1 in 2 people they spoke to said their mental health had suffered during the pandemic, with those anticipating ongoing financial issues being the most likely to cite that their mental health had suffered. Add the documented stress around the risks of contracting the virus (Cavazzoni et al. 2022; Yıldırım et al. 2022), and the complex health issues begin to multiply.

5.4.1 Governments to the Rescue

The COVID period brought more bad news for those struggling to afford a decent roof over their head. It raised many risks, including the increased risk of eviction, especially as the rental affordability crisis threatened to worsen. The concerns among people at the most disadvantaged end of the private rental sector were palpable. Tenants were worried about eviction, especially when the economic slowdown began to hit and many workers lost work hours or were made redundant or couldn't effectively work due to the risk COVID posed to their health. Tenant and other support agencies reported that people were already being evicted from the very beginning of the public health lockdown measures. A blog by the Tenants Union of New South Wales (Mowbray 2020) related several stories.

> We heard from Amrita who has just lost all her casual shifts, and her husband who had to give up driving a cab because he falls into a 'high risk' category. When they reached out to their landlord to ask about reducing their rent last week, they mentioned some repairs the house needed due to recent damage from heavy rains. Instead of rent relief their landlord immediately sent them a 'no grounds' eviction (para 10).
>
> Kye told us she'd been renting her home along with her five-year-old daughter for nearly two years. Two weeks ago she received a 'no grounds' eviction. She's really worried. "I can't afford to move," she said. "And have you ever tried moving with a five year old under your feet? I don't know how we'll find something decent in the middle of this crisis. I've faced homelessness before. I do NOT want to face it again" (para 11–12).

Stories like these were repeated in cities around the country and threatened to become a common COVID-related storyline.

Although initially slow to act, the federal and state governments began to acknowledge the impacts of the loss of employment and earnings for low-income private renters. Within the government policy responses was the introduction of measures to reduce eviction risks for those impacted by the pandemic-induced economic slowdown. Indirect measures such as temporarily increasing income support via a $550 per fortnight supplement for people on JobSeeker payments or the introduction of JobKeeper payments for eligible workers who had lost their jobs provided a boost for vulnerable renters. Still, a more direct measure was the introduction of a moratorium on evictions. All jurisdictions introduced legislation banning evictions for those impacted by the economic fallout from COVID-19. The general intent was to protect vulnerable renters from being kicked out of their homes. To be eligible, a person had to be suffering COVID hardship, the definition which included a person:

- *suffering COVID-19 or has someone in their care who is suffering from COVID-19;*
- *in quarantine resulting from a direction;*
- *unable to work because their place of employment closed, or loss of business due to public health direction (including loss of trade);*
- *self-isolating because they are vulnerable, live with or are a primary carer for a vulnerable person;*

- *restricted on travel so they cannot work or return home due to public health order, or unable to leave or return to Australia.*

and

- *has lost 25% or more of income, or the rent is 30% or more of income.*
- *if there's more than one tenant or resident on the agreement, the 25% loss of income as well as the 30% of income paid on rent, noted above, applies to combined household incomes (Tenants* Queensland 2020*).*

Surveys widely recognised that providing these income measures and other interventions, including the eviction moratorium, significantly improved the well-being of individuals and families eligible to receive them, specifically those in vulnerable private rental situations. In one such survey (Wilson et al. 2021), it was reported that 80% of the survey respondents experienced life becoming a bit or a lot easier as a result of the interventions. For those receiving the increased Jobseeker payments, 51% stated that it made life a lot easier, while 18% said it made life a bit easier. Similar sentiments were expressed for those who received the government's Jobkeeper payment with 53 and 35%. The impacts of the delays in evictions were significant. A full 71% of the survey respondents stated that removing the risk of evictions had made life a lot easier (Fig. 5.7).

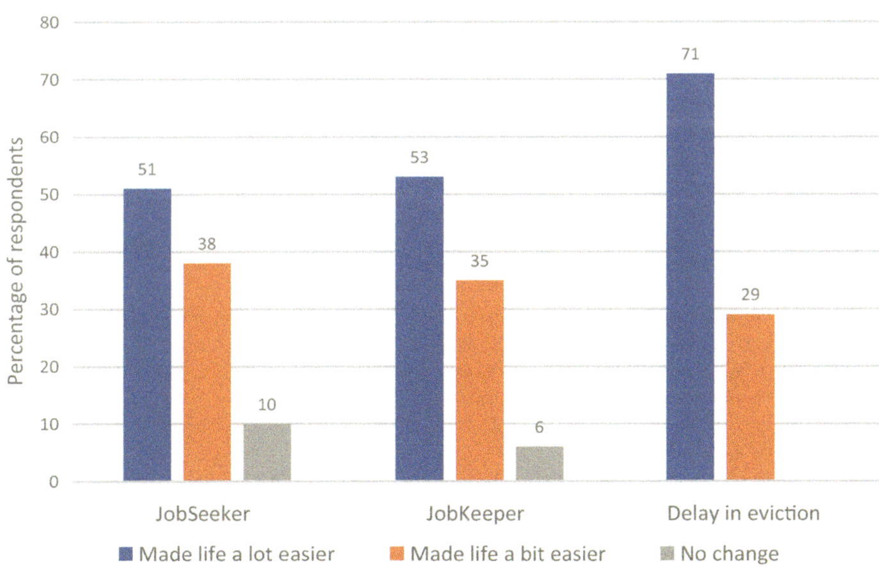

Fig. 5.7 Impact of type of support/benefit (Wilson et al. 2021)

5.5 Housing and the Post-pandemic City

Where to for Australian housing? Like many issues discussed in this book, the housing crisis across Australian cities was evident much earlier than when COVID-19 became a global issue. Cities were the home to significant numbers of people living rough or in other forms of marginal housing. There was also no shortage of private rental tenants suffering stress related to rising rents and falling incomes. The pandemic just moved these issues into the public eye a bit more, albeit for a short time. Despite announcements that we are now 'living with COVID', code for we have to get back to business as usual, the full effects of the pandemic are still likely to be being played out. This will mean that the bumpy ride for those living in the most disadvantaged housing situations within Australian cities will not be over, and the pressures bought about by rental housing costs will not let up. This is despite government proclamations that, economically and in public health terms, we are over the worst of it.

References

Australian Bureau of Statistics (2012) Information paper—methodology for estimating homelessness from the census of population and housing. Retrieved from https://www.abs.gov.au/ausstats/abs@.nsf/Latestproducts/FF268318B1601D62CA257A6F0012ACC1?opendocument

Australian Bureau of Statistics (2018) Census of population and housing: estimating homelessness. Retrieved from https://www.abs.gov.au/statistics/people/housing/census-population-and-housing-estimating-homelessness/latest-release

Australian Bureau of Statistics (2022) Building Activity Australia, December 2021. Retrieved from https://www.abs.gov.au/statistics/industry/building-and-construction/building-activity-australia/latest-release

Australian Institute of Health and Welfare (2021a) Housing affordability. Retrieved from https://www.aihw.gov.au/reports/australias-welfare/housing-affordability

Australian Institute of Health and Welfare (2021b) Older Australians. Retrieved from https://www.aihw.gov.au/reports/older-people/older-australians/contents/housing-and-living-arrangements

Baker E (2020) Homelessness is hard enough, then coronavirus makes life on the streets even more dangerous. ABC News. Retrieved from https://www.abc.net.au/news/2020-04-24/coronavirus-making-being-homeless-even-more-dangerous/12178886

Baker E, Daniel L (eds) (2020) Rental insights: a COVID-19 collection. The Australian Housing and Urban Research Institute, Melbourne

Baker E, Bentley R, Beer A, Daniel L (2020) Renting in the time of COVID-19: understanding the impacts. Retrieved from https://www.ahuri.edu.au/research/final-reports/340

Bonnefoy X (2007) Inadequate housing and health: an overview. Int J Environ Pollut 30(3–4):411–429

Buckle C, Gurran N, Phibbs P, Harris P, Lea T, Shrivastava R (2020) Marginal housing during COVID-19. Retrieved from Australian Housing and Urban Research Instittute Limited, Melbourne: https://www.ahuri.edu.au/research/finalreports/348

Cavazzoni F, Pancake R, Veronese G (2022) Impact of COVID-19 pandemic on mental health and quality of life. An exploratory study during the first outbreak in Italy. Psychological reports, 00332941211066259

Coates D, Anand P, Norris M (2015) Housing, happiness and capabilities: a summary of the international evidence and models

Convery S (2021) Rental stress on the rise as low-income Australians priced out of market. The Guardian. Retrieved from https://www.theguardian.com/australia-news/2021/nov/24/rental-stress-on-the-rise-as-low-income-australians-priced-out-of-market

Crouzet L, Scarlett H, Colleville A-C, Pourtau L, Melchior M, Ducarroz S (2022) Impact of the COVID-19 pandemic on vulnerable groups, including homeless persons and migrants, in France: a qualitative study. Prev Med Rep 26:101727

Hashim MJ, Alsuwaidi AR, Khan G (2020) Population risk factors for COVID-19 mortality in 93 countries. J Epidemiol Glob Heal 10(3):204

Hu M, Roberts JD, Azevedo GP, Milner D (2021) The role of built and social environmental factors in Covid-19 transmission: a look at America's capital city. Sustain Cities Soc 65:102580

Kang S-J, Jung SI (2020) Age-related morbidity and mortality among patients with COVID-19. Infect Chemother 52(2):154

Melbourne Institute: Applied Economic & Social Research (2020) Taking the pulse of the nation. Retrieved from https://melbourneinstitute.unimelb.edu.au/news/news/three-in-ten-financially-stressed-over-impact-of-covid-19

Mowbray J (2020) Uncertainty and fear: evictions during the pandemic. Retrieved from https://www.tenants.org.au/blog/uncertainty-and-fear-evictions-during-pandemic

Owen E (2022) Two years on: six ways COVID-19 has shaped the housing market. Retrieved from https://www.corelogic.com.au/news-research/news/2022/two-years-on-six-ways-covid-19-has-shaped-the-housing-market#:~:text=Housing%20debt%20levels%20hit%20record%20highs.&text=As%20of%20January%2C%20total%20outstanding,from%20139.2%25%20in%20March%202020

Patel J, Nielsen F, Badiani A, Assi S, Unadkat V, Patel B et al (2020) Poverty, inequality and COVID-19: the forgotten vulnerable. Public Heal 183:110

Pawson H, Martin C, Sisson A, Thompson S, Fitzpatrick S, Marsh A (2021) COVID-19: rental housing and homelessness impacts—an initial analysis

Redman E (2021) Victorian renters struggle to afford food and risk eviction, a year into COVID-19 crisis. Domain. Retrieved from https://www.domain.com.au/news/victorian-renters-struggle-to-afford-food-and-risk-eviction-a-year-into-covid-19-crisis-report-1038371/

SGS Economics and Planning (2020) Rental affordability index, December 2020 key findings. Retrieved from https://shelter.org.au/site/wp-content/uploads/RAI-2020-FINAL_compressed.pdf

Tenants Queensland (2020) The evictions moratorium, what it means and our thoughts. Retrieved from https://tenantsqld.org.au/the-evictions-moratorium-what-it-means-and-our-thoughts/

Tinson A, Clair A (2020) Better housing is crucial for our health and the COVID-19 recovery. Heal Found 1–25.

Varshney K, Glodjo T, Adalbert J (2022) Overcrowded housing increases risk for COVID-19 mortality: an ecological study. BMC Res Notes 15(1):1–4

Wilson E, Sama M, Johnson B (2021) No fighting chance: impact of the withdrawal of COVID-19 income and tenancy benefits. Retrieved from Melbourne: https://www.unitingvictas.org.au/wp-content/uploads/No-Fighting-Chance-Final-Report.pdf

World Health Organization (2018) WHO housing and health guidelines

Yıldırım M, Akgül Ö, Geçer E (2022) The effect of COVID-19 anxiety on general health: the role of COVID-19 coping. Int J Ment Heal Addict 20(2):1110–1121

Chapter 6
Moving Around the COVID City

6.1 COVID-19 and the Transport Reshuffle

Just like the changes to many other features of city life, the arrival of COVID-19 into Australia heralded a change in the ways in which people moved around cities. Driven by a range of factors including public health restrictions that were introduced to 'stop the spread' of the virus and long periods of lockdown which anchored people in their homes and local communities, how we travelled for work, shopping and play shifted.

Of course, the COVID-19 pandemic hasn't been the only crisis that has impacted the travelling publics' behaviour. In recent memory, we have had a number of health crises such as SARS and the Avian Flu outbreaks of 2005 and 2013 all had some impact on travel decisions in the countries most affected. For example, the Severe Acute Respiratory Syndrome (SARS) crisis of 2003 resulted in a reduced demand for air travel in the affected regions by about 35% during the height of the crisis. Even during the post-crisis period demand continued to be below the business-as-usual level for six months, resulting in an 8% annual decrease in revenue. The SARS crisis also affected non-essential trips using public transport. In Taipei, for example, there was a reduction in underground railway trips by 50% during the peak of the SARS epidemic. Moreover, it took up to four months for passenger numbers to return (Sung and Monschauer 2020).

The difference with COVID was the scale of the impact. Very early in the life of the pandemic, travel by city dwellers (and others) in Australia fell off a cliff. As mitigation measures began to increase the amount that people could or were willing to move around decline significantly. We noted in Chapter one the impact that COVID restrictions had on the central business districts of Australian cities, turning them into virtual ghost towns overnight. But it wasn't just the CBDs that saw a reduction in travel. It was happening all over. An early review of travel behaviour by researchers at the University of Sydney (Beck and Hensher 2020) using the CityMapper mobility index showed sharp drop-offs in public transport, walking and cycling in Sydney and Melbourne over the first two months of the pandemic (Fig. 6.1). The same researchers

S. Baum et al., *Pandemic Cities*, Cities Research Series,
https://doi.org/10.1007/978-981-19-5884-7_6

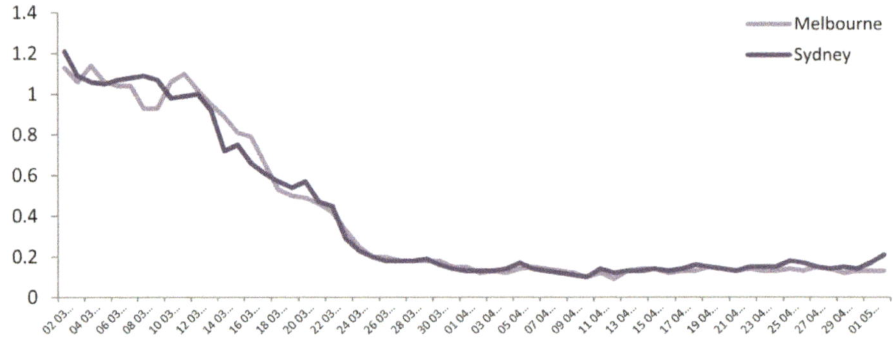

Fig. 6.1 CityMapper mobility index (Beck and Hensher 2020)

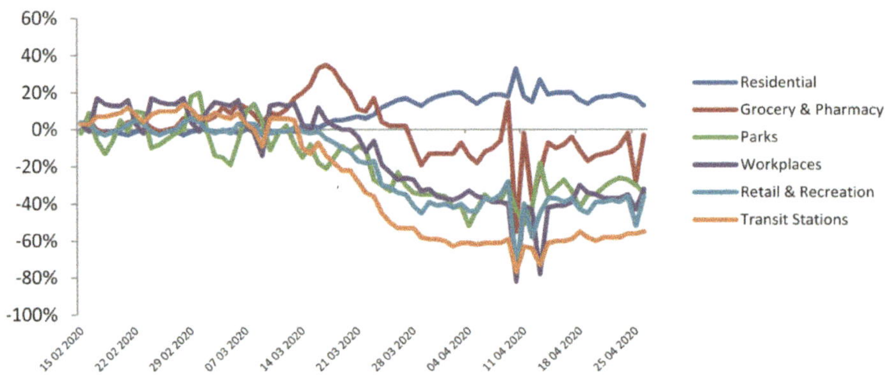

Fig. 6.2 Google mobility report for the whole of Australia (Beck and Hensher 2020)

also showed that people were observing public health warnings to stay at home, and visits to parks, workplaces, retail and recreation outlets and transit stations all fell (Fig. 6.2).

Like other aspects of the COVID puzzle, the great transport reshuffle reflected a broader set of social factors. While overall travel was down, especially during the early days of the pandemic, it was also true that spatial inequalities influenced mobility patterns during the major lockdowns (Fig. 6.3). These graphs are based on mobility data provided by Facebook, which provides a useful relative measure of variation in mobility. The data reveal differences in mobility in Melbourne in areas of high and low socio-economic status. During the early phases of the pandemic in 2020 and into the long Melbourne lockdown from late July 2021, mobility in high socio-economic suburbs was lower than in areas of lower status. The pattern in the long lockdowns in the second half of 2021 showed a reversal of this pattern, with more movement in higher status suburbs. A full understanding of these variations would require a more detailed analysis than we have been able to undertake. Even so, it would be reasonable to speculate on the differences in the 2020 data being

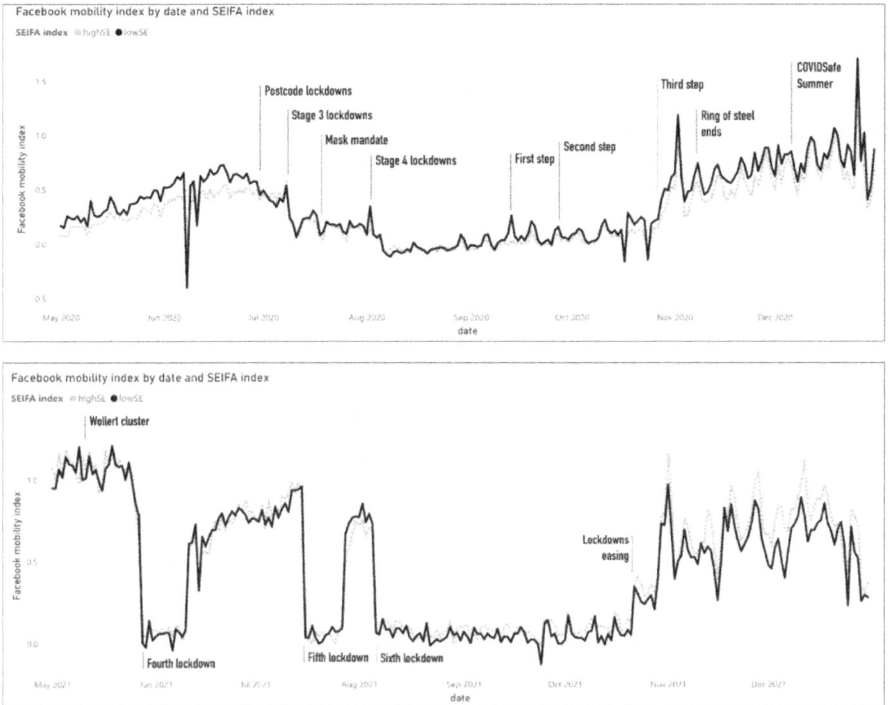

Fig. 6.3 Facebook mobility data by socio-economic status of neighbourhood, Melbourne (Based on analysis by Rohan Byrne,[1] THUD Research Hub, MSD, University of Melbourne)

linked to differences in opportunities to work from home and in non-discretionary travel needs such as care for ageing family members. In 2021, the concentration of COVID cases in the north and west of Melbourne led to a greater fear of contracting the virus—and the focus of policing of lockdown restrictions was also concentrated in these suburbs (Silva 2020).

The lifting of severe lockdown measures at various stages in different jurisdictions did result in a freeing up of travel behaviour, but in many ways, the potential impact of the lockdowns and the concern surrounding COVID-19 and its seemingly multiple variants saw changes that had the potential to be much longer in duration than just simply the early stages of the crisis. Changes in the way we worked, shopped, played and socialised saw Australian city dwellers fall out of love with public transport, become more hooked on the private car and become more active.

[1] Facebook, which, despite the obvious limitations due to the sample being limited to Facebook users with their phones turned on, provides a useful relative measure of variation in mobility by all modes over the course of the pandemic, including some 'noise' on a weekly cycle due to differences in the motivation for travel on weekdays and at weekends.

6.2 We Fell Out of Love with Buses (and Other Public Transport)

Increasingly in recent years, public transport in Australian cities has been, to a greater or lesser degree, planned to replicate the successes of cities in Europe and Canada by creating connected networks that allow more people to reach their destinations with relative ease (Mees 2010). Public transport based on such networks is good for everyone. It reduces congestion on the roads by removing private cars, it reduces greenhouse emissions and politicians love to announce brand-new shiny pieces of transport infrastructure that promise to do wonders for a city or region. As a result, public transport use was growing in most cities.

But with the arrival of COVID-19, we fell out of love with those shiny pieces of public transport. If we didn't need to catch a bus we didn't, we looked for other alternatives. The same thing with trains and other modes of transport labelled 'public'. The extent of the fall was different in different places, but it was obvious to those looking at transport patterns, that public transport patronage fell faster and further than car travel during the early days of the pandemic. Australian news reports characterised public transport as getting the 'cold shoulder' amidst COVID fears and quoted commuters shifting to cars (including lengthy car commutes) because "family has to come first in these situations" and "the upside is I don't have to use public transport and potentially catch COVID" (Fowler 2022). Anecdotally, for those who were out and about, it must have seemed that being a bus driver was a particularly lonely experience with practically empty buses plying what were once busy transit routes.

The available research provided often stark illustrations of the extent of the COVID-led demise in public transport use. Using Apple mobility data, transport economist Mathew Beck and his colleagues (2021) showed that public transport use fell by 80% in Australia during the first wave of the pandemic. An Australian Bureau of Statistics Survey in March 2021 found that around one in seven (14%) Australians reported using public transport one or more times a week in March 2021, compared to almost one in four (23%) who reported using public transport before the introduction of public health measures in March 2020 (Fig. 6.4) (Australian Bureau of Statistics 2021). When the Bureau dug deeper, they also discovered that out of those who regularly used public transport before COVID hit, around 18% had not used public transport since March 2020. People were also asked about their future intentions regarding public transport use. A large group of people (61%) expected to see their use of public transport to remain the same, while some (13%) thought it would increase. A small 7% thought their public transport use would fall, while the remainder were unsure (10%) or said they would not use public transport in the future (10%).

There were clear reasons why we fell out of love with public transport in such great numbers. Obviously, the public health orders that required us to stay at home during the early phases of the pandemic meant that people just stopped travelling to work and school as well as other places that required a public transport trip. The demise in central business districts (see Chap. 1) during the early stages meant that

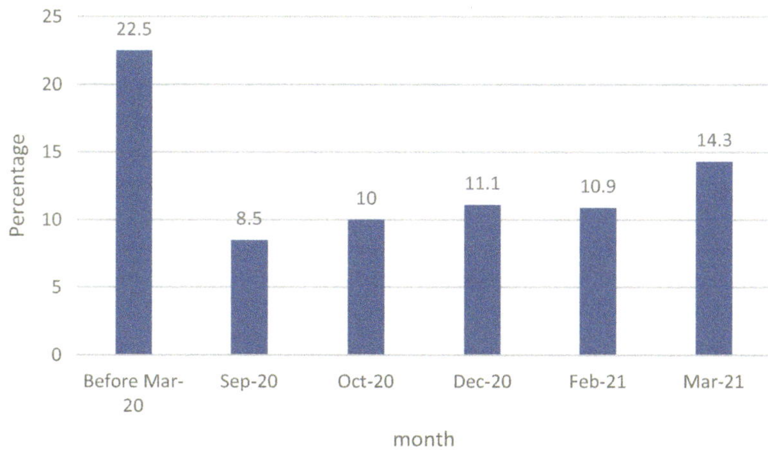

Fig. 6.4 Persons aged 18 years and over use public transport one or more times a week, by the time of reporting (Australian Bureau of Statistics 2021)

buses, trains and trams that were once heaving with commuters on their daily trip to work became, like the destinations they travelled to, empty shells of their former self.

But there were other reasons for the public transport demise, especially once people started moving around the cities again. We were told time and time again that close contact was a key enabler of viral spread. Public transport authorities started blocking off seats to ensure a safe distance was maintained, with the immediate effect of reducing public transport capacity. Less capacity translates into lower usage. But this is only part of the story. One-point-five metres became the magic number if one wanted to be safe, and to this end, public transport was demonised as being a potential hotspot for viruses. This shift in preferences has been labelled 'dread behaviour' by some (Sung and Monschauer 2020), but the COVID risk status of public transport does not appear to be proportional to the actual risk. The Organisation for Economic Co-operation and Development (OECD) has noted, for example, that there were very few clusters linked to public transport during the periods of tight lockdowns (OECD 2020). But despite this, the view of public transport as being equivalent to COVID risk has reshaped the mobility landscape.

Although the Australian Bureau of Statistics tells us that most people are comfortable using public transport, the number is down from what it was in pre-pandemic days. The COVID risk posed by hopping on a bus was too much for many. An ongoing survey of transport behaviour run by The Institute of Transport and Logistics Studies (*ITLS*) at Sydney University found that anxieties about hygiene and crowding levels have diminished over time but are still higher than at the start of the pandemic. A reduction in gender and age differences in responses was also observed, with all groups sharing similar levels of concern (Beck et al. 2021). We saw similar findings in an internationally reported study of 15 countries including Australia (Abdullah et al. 2020) where the authors investigated a myriad of different factors which affect travel choices prior to and during the COVID-19 pandemic. The factors included traditional factors, things like cost, service quality and travel time, but also included a range of health-related factors such as social distancing, infection concern and cleanliness. For the majority of people, factors relating to passenger health risk were considered a high priority during COVID-19, while only a small number of participants thought that traditional factors like cost, travel-time savings and comfort had a similar high priority. For example, before the pandemic hit, cleanliness was judged a high priority by only 34% of travellers, compared to 62% of travellers during the pandemic.

6.3 We Retreated to the 'Safety' of Our Cars

Since the 1950s, Australian cities were built for private cars. Cars were a status symbol and roads and freeways were an indicator of progress. But with the arrival of COVID, private cars took on another symbol, one of safety. Like nearly all modes of transport, the beginning of the pandemic and the public health lockdowns saw travel using private cars fall. Indeed, as has been pointed out by researchers, the largest fall in aggregate trips was by those undertaken in private cars with a fall of around 50% (Beck and Hensher 2020).

However, this was just the beginning of the pandemic, and once stay-at-home orders were eased, private vehicle use got back on track. While public transport was demonised, a private vehicle was considered to be a safe haven away from all the nasty COVID germs. Transport planners, who long have called for more public transport use to improve our cities, wrote about "Cars rule as coronavirus shakes up travel trends in our cities" (Sipe 2020) and looked on in wonder as city dwellers fell more in love with their cars. One such planner was Neil Sipe who looked at changes in transport choices between mid-January and mid-July 2020 noting that, besides Melbourne which was still in the grips of a prolonged lockdown, other large Australian cities saw a growth in private car use (Fig. 6.5).

Journalist writing about the come-back of the car expressed statements like "Rumors of the Demise of Cars Have Been Greatly Exaggerated" (Cadman et al. 2021) and wrote passages such as.

> Just a few years ago, obituaries were being written for the motor car — Millennials and Gen-Zers, so the argument went, were going to swerve away from car ownership as more of them moved to cities with myriad public transport options and ride hailing services like Uber. Besides, concern about the adverse impact on the environment would deter young people from making purchases (para 3).

They noted that the arrival of the pandemic had changed everything.

Motivated by fears of contagion, the rapid growth in car use occurred at a time when large numbers of Australians living in cities were spending some of their days working from home and we had seen a significant upswing in the use of online shopping and home delivery services. More Uber-eats meant less people nipping down

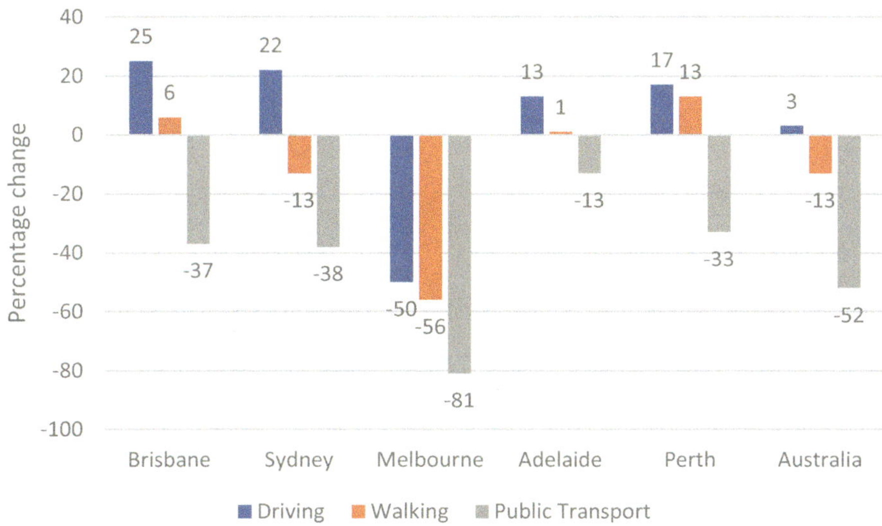

Fig. 6.5 Changes in driving, walking and public transport, January 13 to July 17, 2020 (Sipe 2020)

to the local takeaway—in the opening phases of the pandemic, it was estimated that online mentions of "Uber Eats" on social media platforms were up 770% (Murphy 2020) and major supermarkets saw an upswing in home delivery ordering.

What about people's intentions after COVID fades? Not surprisingly, surveys suggest that we will still see our cars as a COVID-safe haven. Toll-road giant Transurban, not surprisingly, claimed that that for residents of Sydney, Melbourne and Brisbane, private vehicles were still going to be the favoured mode (Fig. 6.6), a finding supported by the survey work of Beck and Hensher (2020) who found that car usage had bounced back to above pre-covid levels once large-scale lockdowns had ended.

The switch to increased private vehicle use and the shunning of public transport had the unintended consequence of adding fuel to an already hot market for second-hand cars. Combined with global shortages of new cars and delivery delays caused by lockdowns impacting on global manufacturing, the market for used cars in Australia reached significant new highs. We all heard stories about people who were able to sell their old second-hand cars for more than what they paid for them. One Guardian newspaper report (Kurmelovs 2022) made the observation that.

> We've always been told cars devalue rapidly, but the booming second-hand car market shows this may not be true in a pandemic.
>
> It was once a rule as sure as gravity: buy a 2012 Toyota in January and it would be worth significantly less by December.
>
> But if you've bought a second-hand car anytime over the last two years, it should be pretty obvious that rule hasn't held true during the pandemic as prices have been driven up by 50% on some models in Australia and overseas.

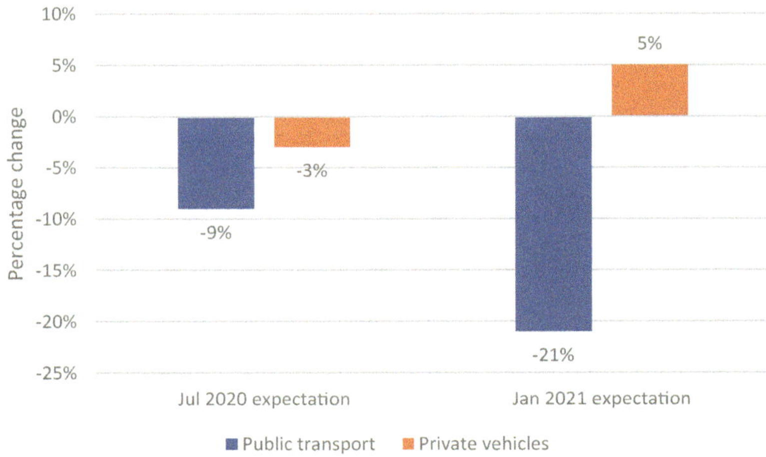

Fig. 6.6 Transport use expectations, Sydney, Melbourne and Brisbane average (Transurban 2021)

> In the small car market, a $10,000 2011 Mazda3 now costs about $12,000, which is more modest but still large, particularly for families looking to buy a new driver their first car. While a Mazda3 that cost $26,000 when new in mid-2020 can be bought second hand for about $34,000.

Research by Infrastructure Australia (Infrastructure Australia 2020) provided a quantitative illustration of the shift in used car demand. Using data from a large second-hand car auction house, they identified a significant decline in the stock of used cars after about mid-2020, just as many Australians were leaving the harshest of initial lockdowns (Fig. 6.7). In true economics 101 fashion, as the supply of used cars dropped, the prices shot up, with close to a 30% increase between the middle and the end of 2020 (Fig. 6.8).

Even as we write, in early 2022, car yards appear empty as stock levels have yet to be replenished, and sticker prices even for the most basic cars seem exorbitant.

6.4 We Got Active

An unintended, but positive outcome of COVID was the increase in residents of Australian cities undertaking active travel. Lockdowns saw people out walking in their local areas, and for communities lucky enough to have them, local parks and bike tracks were teeming with people out on two wheels. The shift to active travel was seen as a good thing. One report in the British Journal of Sports Medicine (Brooks et al. 2021) pointed to the positive impact the COVID-induced increase in active travel could have on reducing the incidence of cardiovascular disease and cancer and improving mental health by off-setting the negative mental health impacts of lockdowns. And this was on top of the potential environmental benefits often ascribed

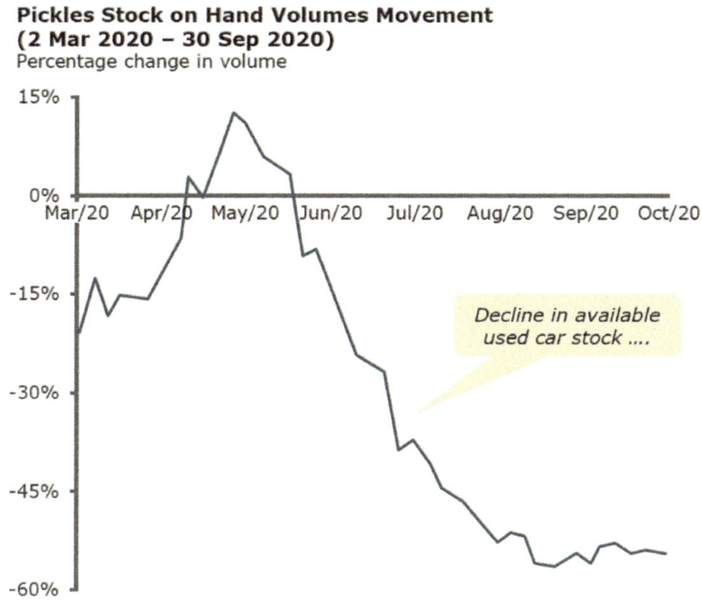

Fig. 6.7 Used car stock levels, Pickles car auctions (Infrastructure Australia 2020)

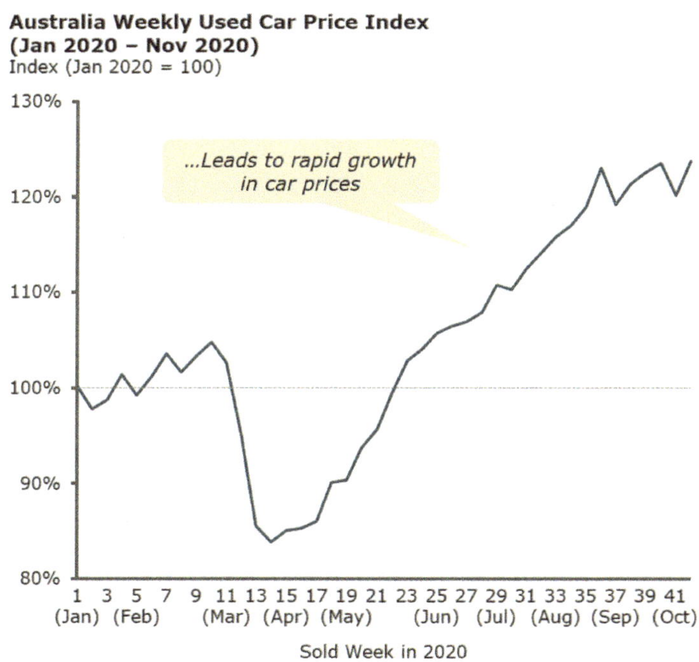

Fig. 6.8 Change in used car prices, Australia, Jan to Nov 2020 (Infrastructure Australia 2020)

to active travel. The irony of the shift to more active travel modes is that for many years active travel has been encouraged because of its potential health benefits, and now it is a global health pandemic that has seen a shift.

Anecdotal evidence suggests that more people walked and cycled to meet their needs for access to goods and services, and for social interactions, especially as local community facilities such as local shops became favoured over CBD localities or other large regional shopping areas. As retail researcher Gary Mortimer (Mortimer et al. 2020) wrote during the pandemic.

> With many more people working from home during the pandemic there has been something of a retail inversion with more people shopping locally. There are clear signs of a resurgence in local shopping villages and high street retailing. There even appears to be a corner store revival of sorts (para 22).

So as local retailing became more popular, more people were riding or walking to take advantage of revamped local facilities. This is in addition to a shift to walking and cycling for exercise and to escape the confines of household lockdown. In one Brisbane newspaper (Stone 2020), a headline read "Cycling spikes on Brisbane bikeways during pandemic" with the accompanying story relating Brisbane City Council data showing that "Across the city there's been a 22% increase in weekday cyclists, and a 91% increase over the weekend" (para 4). The same article made reference to the spike in bicycle sales suggesting one Brisbane bike store had reported a 300% increase in the sale of bicycle helmets, and children's bike sales were up 400% (para 7).

For those who frequented bike shops the evidence of the increase in the popularity of the humble bicycle was clear. Stores no longer had a stock of shiny new bikes with stories of projected waiting times for some models blowing out to 12 months or more. Analysing industry data, one particular website proclaimed:

> It's Official – 2020/21 was a Record Year for Bicycle Imports", showing that within Australia the 2020/21 financial year broke the all-time record for bicycle imports with a total of 1.69 million bikes coming into the country, a 44% increase from the previous year (Latz 2021).

It is unclear how much extra walking people did, but we can get an idea of the upswing in cycling. In a paper published in the Journal of Transport Geography, researchers Glen Fuller and his colleagues (2021) identified that in Sydney, the COVID lockdowns and accompanying risk-related behavioural changes saw large numbers of people (63%) increasing their cycling activity during the height of the public health restrictions. What they also discovered was that a large number of people who were new to cycling (55%) saw their cycling increase even after the restrictions had been relaxed. Another 28% reported that their activity stayed the same. In comparison to this group, 51% of those who had always cycled said that their cycling activity was unchanged from the restriction heights, while another 35% thought they had increased their activity (Fig. 6.9).

One of Australia's peak cycling groups, Bicycle Network Australia, surveyed 1000 people to find out about their cycling intentions and found that 73% of those they spoke to thought they would ride their bike more than usual. Given that more

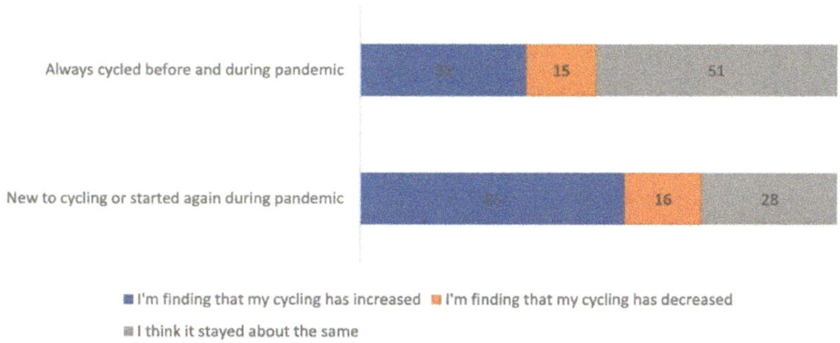

Fig. 6.9 Cycling activity after restrictions relaxed by the history of cycling (%) (Fuller et al. 2021)

than 3.4 million Australians ride a bike every week, the potential to increase active travel was significant. The research also identified that a large number of people were planning to use a bicycle to access essential services such as shops, healthcare and places of work and education, and many were using their bicycles to get their daily exercise during lockdown (Bicycle Network Australia 2020).

6.5 After the Reshuffle

So, was the COVID-19 pandemic a mere speed bump for transport in Australian cities? In many ways, this question is still being played out, but there are some signs that the crisis may lead to some changes in the way people choose to travel around cities. Previous crises, including contemporary pandemics like SARS, have shown that people often turn to new transport practices immediately following a crisis. Sometimes, these are permanent, but other times there is a slow drift back to pre-crisis usage patterns (Sung and Monschauer 2020).

We have already seen that people's attitudes towards public transport seem to be shaped by the COVID risk threat, and for some, this may remain an ongoing concern. Transport economist Mathew Beck and his colleagues (Beck et al. 2021) in the research referred to earlier in the chapter noted that post-pandemic public transport behaviours were found to be dependent on how people used transit before COVID. Transit use has stayed more consistent for people with positive views of the system before COVID, while those with previous negative views have used it less. They found that while car use quickly returned to or exceeded pre-COVID levels, public transport trips stayed at around 60%.

As the economy began to re-open and workers gradually returned to workplaces, overall travel increased. However, continuing patterns of working from home, and associated shifts to more localised shopping and recreation, have meant that instead of the usual assumptions about the direction and scale of peak hour travel that have driven transport engineering practice for many decades, a new set of rules have come into play. Weekday traffic flows are closer to the dispersed patterns previously only seen at weekends. Without the normal 'rush hour', it is not so easy to optimise flows through an intersection. Thus, car-based congestion has spread rapidly in all directions, even with lower absolute traffic volumes.

To make things worse, the slow return to public transport has caused more problems in cities with centres traditionally served by public transport. Modelling by researchers at Vanderbilt University (Hu et al. 2020) in the US showed an 85% shift of public transport users to cars would increase daily commute times significantly. In Sydney and Melbourne, with car-dominated suburbs but transit-dominated centres, a shift to cars for CBD trips very quickly overwhelmed the capacity of suburban road networks. One newspaper headline in early 2022 read "Traffic jams are back as commuters shun public transport" (Waters and Butt 2022), with the reporters talking about 'massive snarl-ups' and 'huge delays' even though the Melbourne CBD's pre-pandemic office spaces remained largely underutilised.

With congestion growing fast and new travel patterns emerging, one might hope that governments would take advantage of the COVID-led pause in rampant population growth to re-direct transport resources in ways that relieve social inequality and confront the climate emergency. Sadly, as we see in the next chapter, governments and industries are doubling down on many of the projects and processes that many planners and communities had been resisting before the pandemic.

References

Abdullah M, Dias C, Muley D, Shahin M (2020) Exploring the impacts of COVID-19 on travel behavior and mode preferences. Transp Res Interdisc Perspect 8:100255

Australian Bureau of Statistics (2021) Household impacts of COVID-19 survey, March 2021. https://www.abs.gov.au/statistics/people/people-and-communities/household-impacts-covid-19-survey/mar-2021

Beck MJ, Hensher DA (2020) Insights into the impact of COVID-19 on household travel and activities in Australia-The early days of easing restrictions. Transp Policy 99:95–119

Beck MJ, Hensher DA, Nelson JD (2021) Public transport trends in Australia during the COVID-19 pandemic: an investigation of the influence of bio-security concerns on trip behaviour. J Transp Geogr 96:103167

Bicycle Network Australia (2020) Australians keen to ride out coronavirus. https://www.bicyclenetwork.com.au/newsroom/2020/04/06/australians-keen-to-ride-out-coronavirus/

Brooks JH, Tingay R, Varney J (2021) Social distancing and COVID-19: an unprecedented active transport public health opportunity, vol 55. BMJ Publishing Group Ltd and British Association of Sport and Exercise Medicine, pp 411–412

Cadman E, Wells C, Hussainpoor S (2021) Rumors of the demise of cars have been greatly exaggerated. Bloomberg. https://www.bloomberg.com/news/articles/2021-07-02/cars-for-sale-surging-demand-in-post-covid-world-makes-prices-expensive

Fowler M (2022) Public transport shunned as drivers rent permanent parking, short-term cars. https://www.theage.com.au/national/victoria/public-transport-gets-cold-shoulder-as-drivers-rent-permanent-parking-short-term-cars-20220119-p59pdv.html

Fuller G, McGuinness K, Waitt G, Buchanan I, Lea T (2021) The reactivated bike: self-reported cycling activity during the 2020 COVID-19 pandemic in Australia. Transp Res Interdisc Perspect 10:100377

Hu Y, Barbour W, Samaranayake S, Work D (2020) The rebound: how Covid-19 could lead to worse traffic. https://lab-work.github.io/therebound/

Infrastructure Australia (2020) Infrastructure beyond COVID-19: a national study on the impacts of the pandemic on Australia. https://www.infrastructureaustralia.gov.au/publications/Infrastructure-beyond-COVID

Kurmelovs R (2022) Secondhand cars: why are they so expensive and when will prices drop? the Guardian. https://www.theguardian.com/australia-news/2022/jan/30/secondhand-cars-why-are-they-so-expensive-and-when-will-prices-drop

Latz P (2021) It's official—2020/21 was a record year for bicycle imports. https://micromobilityreport.com.au/micromobility-products/bikes/its-official-2020-21-was-a-record-year-for-bicycle-imports/

Mees P (2010) Transport for suburbia: beyond the automobile age. Earthscan, London

Mortimer G, Grimmer L, Maginn P (2020) The suburbs are the future of post-COVID retail. The Conversation. https://theconversation.com/the-suburbs-are-the-future-of-post-covid-retail-148802

Murphy P (2020) Australia's changing online behaviour during the coronavirus. AdNews. https://www.adnews.com.au/news/australia-s-changing-online-behaviour-during-the-coronavirus

OECD (2020) City responses. https://www.oecd.org/coronavirus/policy-responses/cities-policy-responses-fd1053ff/

Silva DS (2020) COVID-19 in the public housing towers of Melbourne: upholding social justice when invoking precaution. Aust N Z J Public Health 44(5):430–430. https://doi.org/10.1111/1753-6405.13041

Sipe N (2020) Cars rule as coronavirus shakes up travel trends in our cities. The Conversation

Stone L (2020) Cycling spikes on Brisbane bikeways during pandemic. Brisbane Times. https://www.brisbanetimes.com.au/national/queensland/cycling-spikes-on-brisbane-bikeways-during-pandemic-20200505-p54q1q.html

Sung J, Monschauer Y (2020) Changes in transport behaviour during the Covid-19 crisis: what can we learn from the lessons of the past. https://www.iea.org/articles/changes-in-transport-behaviour-during-the-covid-19-crisis

Transurban (2021) Urban mobility trends from COVID-19. https://www.transurban.com/content/dam/transurban-pdfs/03/Mobility-Trends-Report-1H21.pdf

Waters C, Butt C (2022) Traffic jams are back as commuters shun public transport. The Age. https://www.theage.com.au/national/victoria/traffic-jams-are-back-as-commuters-shun-public-transport-20220303-p5a1ca.html

Chapter 7
Planning the COVID City

7.1 Planning the COVID City

As the previous chapters attest, the COVID-19 pandemic has impacted Australian cities in a multitude of ways and these impacts have had an array of different outcomes. As we have lived through the pandemic, we have witnessed rapid shifts in employment locations and in the patterns and scale of internal and international migration. We have seen the public health emergency reflected in and reinforcing a range of existing inequalities across communities and changing demands for the way space is used within cities. For urban planners, whose jobs demand they anticipate trends and help shape the future of cities, the rapid changes brought about by the health and economic responses to the pandemic raised significant challenges.

Contemporary planning trends valuing high-density living and encouraging the use of public transport instead of private cars were viewed as a danger in the context of a virus which was spread by close person-to-person contact.

Much of the careful research and analysis undertaken by academic and practising planners in Australia takes place at some distance from the calculations of political decision-makers: who are, in the main, locked into growth-based agendas. To try to bridge this gap, planners attempt—not infrequently—to package their agendas to fit the demands of prevailing imperatives, particularly anything that is recognised as a political crisis. These crisis narratives can, in turn, suddenly propel planning projects forward. The current public health crisis has for example offered opportunities to fast-track projects including the creation or extension of bike lanes and the re-purposing of some on-street parking for hospitality and open space. Planning has, however, a complicated history of being shaped by crisis narratives.

S. Baum et al., *Pandemic Cities*, Cities Research Series,
https://doi.org/10.1007/978-981-19-5884-7_7

7.2 'Pop-Up', Fast Track, and Re-opening the Centre to Cars: Australian Planning Reforms in Response to COVID

The rapid changes in behaviour of a huge majority of Australians in response to the COVID public health emergency have shown a willingness to follow leadership when political and 'expert' voices are presenting a united message. Political responses to COVID have given urban planners a lesson in how quickly some urban planning practices can change, but have also demonstrated the weakness of commitment to some key tenets of planning strategies.

Some changes to urban planning policies and practices introduced during the pandemic are notable because they represent a disruption to previously entrenched political and institutional barriers to 'ideal' planning objectives. A greater role for *local* amenities and connections is, for example, consistent with long-standing planning principles under various guises from the twentieth-century 'neighbourhood unit' through to the latest Melbourne terminology, the '20-minute city'. Other changes—like a further increase in private car-based traffic—represent a direct reversal of planning principles around transport mode and, more broadly, equity in planning processes. The extent to which such changes persist in the post-pandemic period will provide important clues to the strength of commitments to these wider objectives and a clue to how planning, at least the form currently conducted by the state, engages with crises.

Here, we consider three strands to planning responses to COVID in Australia:

- The 'pop-up' response;
- Re-emergence of policies for car-based city centres;
- (Further) erosion of 'checks and balances' on political support for major infrastructure.

7.2.1 *'Pop-Up' and Active Transport Responses*

The context and narrative of *crisis* around COVID propelled certain planning projects forward. An emphasis on social distancing, lockdowns limiting movement to local areas and to limited purposes such as exercise and the association of public transport with COVID offered opportunities to fast-track projects to create or extend space for walking and cycling. This included temporary closures of streets or road lanes to cars, as well as reallocation of road space through temporary and more permanent infrastructure. In Victoria, it included a state-sponsored program of 'pop-up' bike lanes along selected major inner metropolitan roads, as well as the rapid procession (termed 'acceleration') of projects for bike lanes in and around the central city which were already included in local (e.g. City of Melbourne) transport strategies. Post-lockdown, these lanes are largely considered to have become permanent: but have also been the subject of conflict over space for car parking, reflecting competing ideas for the reinvention of the central city (Taylor 2020).

With the easing of restrictions, 'pop-up' planning also took the form of repurposing of (some) on-street parking for hospitality and open space to allow businesses to operate under restrictions on density limits and distinctions between indoor and outdoor spaces. 'Parklet' programs were common: the quick conversion of public street parking spaces into spaces oriented towards hospitality and retail venues, sometimes paired with vouchers and other incentives to entice shoppers and diners back to urban areas. At the local level, alliances of businesses and government quickly sought to reimagine (and recapitalise) depleted central cities in a 'COVID-safe' way—including twenty-first-century placemaking techniques such as pop-up hospitality installations for footpath and parking space dining. The rush to save or reinvent inner-city destinations also, and somewhat contradictorily as we note below, borrowed heavily from the free car parking and car-oriented retail renewal that informed mid-twentieth-century urban planning.

Pop-up projects are of course premised on temporality and have been subject to only limited evaluation. They also tend to be spatially uneven: arising less from long-run planning than from a happenstance of existing plans and the interests of locals, including local businesses. This temporary nature is identified as a strength in terms of making change possible and of testing ideas (Rowe 2021). But it also renders their fates fragile: dependent on competing visions for the post-COVID city such as the role of cars versus other forms of transport and uses of space. Gertrude Street in Collingwood, for example, has been championed as a successful conversion of parking space. Other projects have been viewed more critically—both outcomes largely based on anecdotal assessments (see "From pop-ups to permanent, bike lanes create new hurdles as city gradually refills" (Fowler 2022), "No more new bike lanes for CBD after council cops complaints" (Gratz 2022). Bicycle advocates, among others, have called for a more transparent evaluation of pop-up projects. In the US context, a proliferation of pop-up bike, pedestrian and hospitality space was documented across many cities. This trend of COVID pop-up was welcomed as an urban shift but also critically questioned: Badger (2020) for example asked why planning decisions, premised on a kind of 'blank slate' of COVID and with no real public oversight, were able to happen so quickly in cities with so many long-standing, unaddressed inequalities in urban amenity and safety (Badger 2020).

7.2.2 Re-opening the City to (COVID-Safe?) Cars

Pop-up planning policies have not been limited to bike lanes and outdoor bars. The same crisis-led efforts to rescue or reinvent flailing central cities in the wake of COVID have also, and often contradictorily, seen a resurgence in policies attempting to lure back private cars. Central city local governments have offered free parking (e.g. Melbourne's free parking voucher program), removal of parking enforcement and general pro-car policies (Adelaide's "happy drivers' month") in attempts to welcome private cars to central cities. All electoral candidates in the City of Melbourne and Sydney local government elections committed to free parking as part of COVID recoveries.

The notion that free car parking can lure shoppers and life back to the central city seems to owe direct homage to the 1950s 'predict and provide' planning strategies that attempted to reshape cities to fit suburban automobile patterns amidst rapid motorisation (Taylor 2020). Yet prior to COVID, recent decades had seen central cities of major Australian cities systematically prioritising pedestrian, bicycle and public transport traffic and alternative uses of the public realm over that of private cars. The quick shift back to free parking and car luring policies during and since COVID—however temporary—is a reminder of the fragility of planning and transport policies. Pop-ups—both of retailing and of free parking—have occurred largely in inner-city areas where the most progressive, incremental work has been undertaken on shifting road space allocation and transport strategies.

At the local government level in the COVID context, Australian CBDs are concerned about decline and hollowing out: hence a set of plans have sought to revise the centres of major cities, in various ways but often through retailing and car parking. Partly on account of the impacts of digital shopping and of working from home trends, retailing and central cities are desperate to save themselves, and often to do so by appealing to 'COVID-safe' cars and parking policies for them. True, parklets and bike lanes also feature heavily: bearing in mind that the main shift has been *away* from public transport: attempts at reinventing areas involve active transport, and cars, but not public transport. Public transport and public crowds are casually associated with COVID and problematised as such. The suddenness of this shift in attitude through COVID saw quick shifts: Paris has reset to expel cars in favour of bikes while Australian cities, such as Melbourne and Adelaide, reset in part to welcome cars.

Part of this reflects the association of private cars with being safe and equitable transport in the COVID context. Everything about the COVID response—the vaccination and testing clinics, the messaging and the shifts in getting groceries delivered to your boot—has underscored the safety of private cars and of digital technology platforms. This message has been supported both by the government and by advertisers. This seems to have supported a shift from public transport and towards greater car traffic since COVID, with more trips made by car overall despite the popularity of cycling and walking. Digital technology, in part, translates to delivery and often to delivery by private vehicle and by 'pop-up' jobs (Uber Eats, food and delivery models that eliminate the need to either leave your house or to get out of a car).

Unrestricted free parking was associated with COVID safety and fair treatment for health workers—state and local governments both committed to removing parking enforcement following reports of a health worker receiving a parking infringement. Free parking has been concomitant with the valorisation of cars and the increase in car traffic, and the vehicle of COVID safety being the car. Thinking of cars as safe and effective COVID transport has been casually supported by many employers (including universities) making car parking free again as a COVID-safe response. In Melbourne's CBD, the fact car traffic had returned to pre-pandemic levels, even with overall trips to the CBD depleted, was one rationale given for stopping bike lane installation—perversely, because *bike* lanes (not an increase in car traffic despite a reduction in overall visitors) were thought to have increased congestion and impacted traders (Waters 2022).

Reprioritising car access to the CBD also reflects and reinforces a shift in metropolitan dynamics. With the work from home shift among other changes (lockdowns not the least of them), Australian CBDs have lost their unquestioned primacy as a destination. Instead of CBDs casually enjoying centrality, they fight to regain or assert their status as destinations. Some of this has been twenty-first- century placemaking parklet style but equally it has involved a return to encouraging cars, free parking and the golden ticket idea that the one thing stopping people (safely) from visiting a destination is the fear they cannot drive their car or even might be faced with a fee for parking it. The push for more, free, uninhibited car parking has enjoyed considerable political cache, at multiple levels of government. It directly returns to (failed) 1950s visions of trying to remake central cities to accommodate the car. It also undermines the work of transport planners over decades both inside and outside of government and has no basis in any research. Even if your limited goal was to bring shoppers back to the CBD if you were to do that you would be on very shaky ground if you were to assume this would happen through free parking. Thus, the risks in crisis-led planning: change is possible, but the direction is fragile, and research may have an even more marginal role.

7.2.3 'Fast Tracking', 'Resetting' and Continuing Erosion of Public Engagement

For all their claims of responsiveness to community opinion, state governments have steadily stifled debate on the merits of large infrastructure projects. They have done this by increasing ministerial powers to 'call in' projects, narrowing the scope of planning panels and environmental effects inquiries and blatantly either obscuring or manipulating cost–benefit calculations. The Victorian government's creation of a 'Red Tape Commissioner' (www.vic.gov.au/red-tape-unit) is just one clear example of this trend.

Some examples of the ways that an acceleration of centralised power in planning is being linked to COVID recovery include, in Victoria, changes titled "Planning revamp to boost economic recovery and jobs" (Wynne 2020) and in NSW through a "Planning System Acceleration Program" under the auspices of the COVID-19 recovery plan (NSW Government 2020). Similar patterns were seen in Western Australia where legislative planning amendments were "accelerated to support the state's economic recovery from COVID-19" (WA Government 2022).

Jobs, big projects and speed are common themes. New South Wales posited its COVID recovery through major infrastructure—measured in multiple billions, with a "pipeline of $100 billion of infrastructure with a $3 billion accelerator fund". The

NSW changes were to bring "forward immediate reforms to support productivity, investment and jobs during COVID-19 with a Delivery Unit focussed on unblocking programs caught up in the system". Planning reforms were boasted to offer a "33% time saving in rezoning decisions and 25% for larger projects", with projects meeting the definition of state significance to be cut further. Precincts for priority and state funding include Sydney's Tech Central—described as "a home for tech giants".

With the (lack of) foresight characteristic of planning, in November 2020, the Victorian Government announced that "Improving the efficiency of Victoria's planning system will help make sure our state's recovery isn't held back by bureaucratic red tape or unnecessary delays". Planning changes in Victoria were described as a "planning revamp to boost economic recovery and jobs" (Wynne 2020), with changes to "speed up delivery of the vital major project and boosting construction of social and affordable housing—and the jobs that go with them". The Victorian Planning Authority "Streamlining for Growth" program also fast-tracked digital shifts in governance. Subsequent (introduced in late 2021 and 2022) amendments to Victoria's planning system around transport infrastructure have been described as "enabl[ing] more efficient delivery of infrastructure", with "alternative approval pathways" (Ministerial rather than local government authority).

Continuing the theme of jobs and centralisation of decisions, the Victorian Government also introduced a "Development Facilitation Program" specifically to streamline large projects and "to support investment and job growth and drive economic recovery". Projects would need to be 'shovel ready', larger scale and in a list of priority sectors. Facilitation means applications are assessed and fast tracked at the ministerial level (https://www.planning.vic.gov.au/planning-permit-applications/development-facilitation-program).

Rather than seeing COVID as an opportunity to reassess major transport infrastructure projects in the light of the pause in population growth and the growing reality of the climate emergency, governments are doubling down on their retreat from evidence-based decision-making. Political imperatives to 'rescue' the construction industry have accelerated progress on suburban transport mega-projects in road and rail (Victoria's Big Build 2022) that were previously justified by rampant growth projections and are now said to be an essential stimulus to rebuild the economy. It also reflects poorly on planning norms that the main reform consistently sought to planning processes (in a long familiar pattern) (Gurran et al. 2014) is to make them faster, and decision-making more centralised.

The May 2022 election delivered a new national government and demonstrated the strength of public appetite for seriously tackling climate change (local car enthusiasms notwithstanding). But, reliance on big infrastructure to stimulate the post-lockdown economy, without regard for long-term social or environmental costs, appears to be undiminished. The new government's commitment to decarbonising urban transport relies almost entirely on a transition to electric (rather than fossil fuel) private cars with no stated role for active or public transport, or for cities or the planning of them. Critiques of this approach still have a long way to go to gain any political traction.

COVID recovery packages have repeated long-standing practice of stimulating housing construction as a means to weather economic downturns (Taylor & Dalton 2015). In the same construction-jobs spirit as state planning acceleration policies, the Commonwealth offered cash for rapid starts to renovations or extensions. Not surprisingly, the majority of those eligible for this scheme were wealthy homeowners with existing plans and approvals that typically would have been built without the subsidy. These middle-class subsidies only served to fuel price rises for construction trades.

There has also been bipartisan support for variations on first homeowner grants. These policies, which aim to reduce upfront costs of access to the housing markets, are known to entrench existing structures in the housing industry and maintain the existing distribution of wealth (and debt) within it.

With a stronger intent to address social inequality in housing, the Victorian government announced in late 2020 that it would spend $5.3 billion to build more than 12,000 'affordable' and 'social housing' dwellings (Homes Victoria 2022). Again, this was to be achieved through 'streamlined' planning making the Minister the responsible authority for all projects under this umbrella and removing normal objection and appeal rights. This has inevitably led to tension with marginalised local governments, and with community objectors who accuse the minister of 'steamrolling' local planning discourse. The scheme has been welcomed by housing researchers as a welcome start to tackling the huge under-supply of social (Raynor 2020). Even so, criticisms remain over the private-sector-focussed 'partnership' model which, in the past, has led to poorly integrated communities and a loss of public land for negligible gain.

7.3 COVID Responses: Implications for Sustainable Transport and Compact Urban Form

As we saw in Chap. 6, the COVID-19 pandemic has changed how Australians live in (and move out of) cities: the shift to driving more cars more often, spending more time at home, revalorising lower densities and housing types with private open spaces, cutting ties from suburbs to centres, being more digitised by shopping and interacting online, a retreat from and at some stages criminalising of physical public space and a renewed association between density and public space with danger. We have seen how swiftly the prospect of public interaction can be cast from the vaunted realm of glossy placemaking brochures into the territory of fearful public health warnings. And how the seemingly omnipresent pull of the Central Business District can decline (and how swiftly those with an interest in the CBD react, both with old and new playbooks). These all have profound implications for the understanding of Australian urbanism and of what is possible in urban policy.

Some of these shifts directly undermine stated planning strategies, while others almost accidentally achieve long-stated goals. The way that COVID-19 and the governance of it have played out in Australian cities overshadows by orders of magnitude any formal strategic plan or planning mechanism. In some ways, the responses to COVID have reinforced planning orthodoxies; in others, they have reversed some planning ideas and assumptions and in others have given sudden unexpected impetus to ideas that have been half-heartedly floating around in planning documents for decades (such as the move to the regions).

A shift in the standing of public transport quickly had public policy implications—for example "while hailed as an environmental solution to the pollution caused by individual car usage, public transport is not ideal in a pandemic situation" ("How do you build a city for a pandemic"? BBC). Elsewhere ideas about the pandemic and its influence on cities (and how planners should respond) focus on the urban design opportunities of expanding public and green space and scope for walking. Those focussed explicitly on access and equity—"the need to create sustainable cities that promote health"—as an argument to "create more walkable neighbourhoods" do so with a focus on COVID rather than planning and urbanism. Arguments against population density, and for green space and walkability, are part of this discourse: but the same articles note that low-income neighbourhoods "are more likely to rely on public transportation" and not own cars—which can be read as an argument simply for buying more cars. In some ways, these are 'classic' twentieth-century planning and zoning ideas, and a return to them is both an opportunity and threat to underlying planning principles.

Some of this shift seems to have influenced Australian planning professionals. A survey by Bolleter et al. (2021) of Australian planning professionals and their attitude to the planning implications of the COVID-19 pandemic found many were reconsidering city size, density, self-sufficiency and open space and housing design on account of COVID. Within this, several specifically questioned goals to boost public transport and criticised an 'over-dependence' on public transport. Surveyed planners also commonly associated density with spreading COVID and were reappraising high-density models of development in favour of smaller self-contained cities. While some planners supported transport alternatives including bikes and electronic micromobility, they did so in part to reduce public transport use. The attitudes of some planners signal some move away from, or questioning of, the planning orthodoxy which in Australia has long supported boosting higher density housing and public transport use. While critical re-evaluation of policy in response to new evidence or threats has its merits, we argue that casually including a principle of reducing public transport provision—despite the myriad equity and environmental costs this might have—signals some of the complex implications of COVID for planning in Australia.

Calls to abandon or dismantle dense cities in light of COVID are certainly not unique to Australia. Neither are moves away from public transport. Jasiński (2022) argued that worldwide, COVID responses for public health have directly threatened ideas of modern urbanism: including increasing the value of private space and the inequitable distribution of access to private space when "public space can no longer replace it". To Jasinski, "some of the dogmas of contemporary planning, such as the pursuit of high building density, promotion of public transport and the strengthening of the role of direct human contact in public space to ensure safety have all been challenged" (Jasiński 2022).

In Australia, the response to COVID threats was widely adopted but unevenly felt: by suburb, housing type and household background, among other inequalities. Outer urban areas had limited access to quality local public space—a legacy of suburban planning orthodoxy. In the inner city, likewise, the closing off of shared and public spaces: parks, playgrounds, recreation and the public and semi-public 'third space' (retail) realm, had unequally distributed impacts including for people living in apartments with limited private space. Public space in the pandemic was replaced with private space and with virtual space: a constantly updated, home schooling, home working, refresh clicking, Netflix binging and Amazon deliveries reality (for the lucky ones). Exposure both to COVID and to COVID-related policies, including those in the planning and transport realm, has been unequal.

Nonetheless, COVID has offered an opportunity to respond to disruptions, including in transport, with a commitment to sustainability and equity. Jasińksi (2022) contrasts two versions of a post-pandemic city: a negative (suburbanisation, return to private cars and surveillance society) and a positive (multifunctional and compact city, responsible and active transport and smart city). So while the reshaping of planning to crisis in any fashion brings risks (we return to in our conclusions), COVID-prompted 'resets' of urban planning can come in different formats. Scholars point to green-oriented recoveries, particularly those (like Paris) seeing a rapid increase in bicycle space and use. Weighting density and public transport as explanations and as planning responses privilege physical explanations of risk and of responsibility, while to many scholars, inequality rather than urbanism is the fundamental problem that planners (as with political debate more broadly) should be focussed on addressing.

In Table 7.1, we list some of the most important changes in transport and housing in Australian cities due to COVID. For each, we identify the alignment with stated goals of urban strategic plans and compare 'ideal' planning responses with our observations of current practice responses. In Australia, these responses have in many respects supported a return to the encouragement of car use and dispersed suburban development. While there are exceptions, the uneven distribution of 'green' infrastructure such as new protected bike lanes (largely limited to local government areas with pre-existing plans) risks furthering the inequalities that already characterise Australian cities and their planning outcomes.

Focussing on public transport responses, a responsive and equity-based approach would see public transport services need to run at high frequencies for many more hours in the day. Many suburban bus services, particularly direct services along arterial roads, should run much more often than their existing peak offerings. Routes can be tweaked to remove unnecessary detours that lead to slow travel times. Most importantly, we need to rapidly create 'pop-up' dedicated bus lanes (Bliss 2019) across metropolitan areas. These lanes allow buses to avoid being held up (Gahbauer and Matute 2019) by increasing traffic volumes. Although bus lanes may reduce the capacity for private vehicles, when buses run frequently they are a much more

Table 7.1 Australian transport and housing challenges in COVID-19 cities

Change due to COVID	In line with stated plan goals?	Response that would be consistent with stated plan goals	Actual policy or practice response
Decline in PT use and slow recovery as travel restrictions ease	No—reverses stated goals	Restructure services to match local travel demand (as required for '20-minute city')	Public transport services have generally been maintained despite severe falls in revenue, but budget pressure for cuts may build. Few moves to systematically redeploy resources in line with new movement patterns
Increase in active transport (walking and cycling)	Yes—movement towards long-established by unreached goals	Systematic support for continued growth in active travel	Some new bike lanes, typically where local government plans were already in place. Cast as 'pop-up' and supporting COVID health response. No apparent systematic moves
Rapid growth in car use after lockdowns/congestion said to be approaching pre-pandemic levels even while some WFH continues	No—reverses stated goals	Policies to reduce demand for car use	Some reversion to 1960s assumptions to support car use (including free parking) COVID-safe messaging: 'Your car is the safest place to be in public'
Sudden shift from large and predictable directional peak flows due to increase in local and cross-town travel	Mixed—reflects changes consistent with '20-minute city' ideas	Assess changes and re-align infrastructure spending to support	Together with stalled population growth, these trends undermine previous business cases for big road and rail infrastructure projects. But, there is no evidence of these being re-considered. Instead, projects are being accelerated, predicated on 'jobs and growth'

(continued)

Table 7.1 (continued)

Change due to COVID	In line with stated plan goals?	Response that would be consistent with stated plan goals	Actual policy or practice response
Increase in local trips and demand for some local activities	Mixed—reflects changes consistent with '20-minute city' ideas	Support for public and active transport at supporting local trips Support for new work and services opportunities in local centres	Local level emphasis on transferring use of public space from parking to hospitality, but also on free parking and car-based recovery
Economic decline in CBDs due to drop in workers, visitors (and temporary residents)	Potentially positive if increases support elsewhere for polycentric development Problematic at local and 'global city' levels	Support for business to make transitions to new locations Focus on sustainable improvements to amenity, transport, land use for CBDs and other centres	Economic power of CBD re-asserted incentives for workers and visitors to return Successful calls to relax car parking restrictions Free parking, 'happy drivers month'—back to failures of the 60s?
Rise in online shopping/deliveries	Mixed—impacts on traffic demand and curb space management	Alternatives to 'white vans' (mobility hubs, micro-mobility and so on) Clarify and support curb space hierarchy policies	Pressure to remove bike lanes Entrenching gig economy
Spatial inequalities in transport access and mode choice	COVID mobility patterns reveal existing issues	Reaffirm commitments to address spatial inequalities in transport needs and access Accelerate suburban bus network reform	Strategic planning assumptions and major investment directions have not yet outwardly changed
Stalled international migration and student arrivals	Undermines key strategic assumptions	Opportunity to re-consider development accelerated by previous rapid growth	Strategic planning assumptions and major investment directions have not yet outwardly changed

(continued)

efficient use of scarce road space. Faster travel times for public transport would, in turn, mean operators could deliver more frequent services with existing fleets and drivers. Frequent services on these pop-up corridors will provide a critical, time-competitive alternative to driving. Although not without its challenges, implementing a fast and frequent bus network is conceptually straightforward and the cost is modest compared to the congestion impacts it could offset.

Table 7.1 (continued)

Change due to COVID	In line with stated plan goals?	Response that would be consistent with stated plan goals	Actual policy or practice response
Population shift to regions	Yes—long-standing on-paper support for decentralisation	Reconsider costs of metropolitan primacy Support regional areas with planning and services for interconnected regional cities; housing diversity; local work locations; dispersed public services	Transfer of metropolitan property price growth to (some) regional areas Accelerated housing construction in regional cities, largely in suburban style Acceleration of previous projects connecting regions to CBDs: regional areas are still considered as dormitory suburbs
Collapse in apartment market	Reverses stated intensification and affordable housing goals Reveals shortcomings of poor design and speculative development models Undermines some TOD ideas	Address the inequitable access to shared or public space during COVID Prioritise liveability and affordability of housing at all densities	Increased desirability of housing with private entrance/garden: division in markets Collapse in apartment market: pressure to prop up existing practices

This solution would require a nimble and co-operative approach from state and local transport authorities and private operators. But, the obstacles are large. Not only are the processes for changing contracts cumbersome (especially in Melbourne where privatisation has extended further than elsewhere), but governments lack the political will to tackle reforms that would be resisted by both the small but vocal captives of the existing bus services and the drivers and traders who view allocations of road space for traffic and parking as their fundamental human right. This rights-based approach has been bolstered by the association with cars and free car parking (and car lanes) as a COVID public health-based response. And, beyond these obstacles that existed for many years, COVID has gutted public transport budgets. The collapse in fare revenue from public transport in Australian cities has largely been absorbed by state governments, although there was a marked difference in commitments to maintaining service levels. In the early part of the pandemic, services were increased to cater for essential workers in Brisbane, while the greatest cuts to services were seen in Sydney.

Regarding apartments and higher density housing, planners in Australia face the challenge of recognising the legacy of poorly designed apartments. Recasting compact city policy in a more equitable way would require moving away from simple binaries of density and its regulation (either to limit or to promote density (Hurley et al. 2017). This may involve retrofitting interventions and policies to make apartments more sustainable and equitable. Tackling the realities of the compact city and its legacies means moving beyond the density debate and engaging with complex literature on building faults, owners corporation management and affordable housing models. Planners need to make future housing, irrespective of density, more liveable and responsive to residents' needs.

7.4 Planning and *Crisis*: Implications of COVID Responses for Practice and Research

COVID has challenged established planning ideas and orthodoxies. Australia is not unique in this—we refer to Jasiński (2022) who argues that worldwide, COVID responses based on the public health crisis have directly threatened ideas of modern urbanism. To conclude, we now reflect on the broader challenges to urban planning posed by its reshaping to fit crises, the latest of which is COVID.

Planners not infrequently attempt to package their agendas to fit the demands of prevailing imperatives, particularly anything that is recognised as a political crisis. These crisis narratives can, in turn, suddenly propel planning projects forward. We have outlined how the current public health crisis has for example offered opportunities to fast-track projects by creating or extending bike lanes and re-purposing some on-street parking for hospitality and open space. And how at the same time it has added to the valorisation of the private car and of desperate 'free car parking' schemes, and to the justifications for planning ministers to further accelerate the 'fast tracking' of major planning decisions and the erosion of meaningful public consultation or engagement. While arguably benign experiments of 'pop-up' bike lanes and open spaces have benefited from COVID rationales, the political imperative to rescue the construction industry has accelerated progress on dubious transport mega-projects previously justified by growth projections which are now, at the least, questionable.

Planners and planning stories rarely normally make good television—it is complicated, it's long term, the instruments used are subtle and there is rarely an unambiguous good and bad. One of the few planners who does make for good television is Robert Moses, New York Commissioner for Planning and 'despotic city planner' (Wainwright 2017), described as a "tempestuous, arrogant and very effective bureaucrat" (Gutman 2008, p. 534). Moses is featured not only in a musical of his standoffs with the planning canonical Jane Jacobs but in the recent West Side Story remake featuring dramatisation of opposition to Moses' major urban renewal projects in New York.

Broadly speaking, planning and how it is taught have sided with Jane Jacobs, not Robert Moses, in the decades since. One of Jacobs' contributions was to recognise that planning (as a profession in the 1960s), in its confident push to create a better urban environment for people, found a lot of *actual* urban environments and actual people inconvenient and sought to get rid of them. This was explicitly framed around '*urban crisis*' and the need for urban renewal. As in one of the New York urban renewal planners of the time, who Jacobs quoted in Life and Death of American Cities:

> My idea is to go just outside centres of population, pick up cheap land, build a whole community, and entice people into it. Then go back into the cities and tear down whole slums and make parks of them (quoted in Freestone (2000)).

William H. Whyte Jnr, later an influential urban designer, also wrote in despair of the anti-urban effects of 'crisis' and 'renewal' in American cities of the 1960s: noting that "most of the rebuilding underway and in prospect is being designed by people who don't like cities. They do not merely dislike the noise and the dirt and the congestion. They dislike the city's variety and concentration, its tension, its hustle and bustle. The new redevelopment projects will be physically in the city, but in spirit they deny it" (Whyte quoted in Gratz (2022)). The logic at that time among powerful planners was that the commitment to creating new ideal communities also brought with it, concurrently, a casual view that existing cities and urban areas could and should be squashed out (by 'renewal', or large roads, or some combination of the two). Jacobs referred to the "cataclysmic use of money for suburban sprawl, and the concomitant starvation of those parts of the city planning orthodoxy stamped as slums".

Moses and urban renewal encapsulated one of the uneasy legacies of the planning profession that has, many writers attest, seen the planning profession and scholarship side more with Jane Jacobs (a study of urban conditions and politics) than with Moses (knocking things down and rebuilding them). Australian planning scholars generally shared a suspicion and turn away from technocratic planning and confident, top-down modernity. The idea that an expert knew best, that they could and should have the power to reinvent cities and that there was an overriding justification in doing so: this has not aged well in twenty-first-'-century planning thought. Some version of advocating for granular, discursive, incremental urban improvements at the human scale characterises much planning scholarship—if not always planning strategy and practice—in Australia. There has been a more common discourse in planning scholarship of communicative, democratic, radical planning: often by embracing the disagreement that characterises the democratic ideal. Moses, meanwhile, had chided that "the problem of constructing large-scale public works in a crowded urban setting…is one which democracy has not yet solved" (quoted in Wallock (1991)). The continued reliance on major infrastructure, and on expedited processes for these projects, suggests the urban political realm in Australia still grapples with the competing urges of doing 'big, bold' interventions on the one hand and of evaluation and democratic oversight on the other.

Urban clearances and renewal projects of the twentieth century were justified and propelled by real and narrative crises. These cast a cautionary note around responding to the crisis in urban planning. A 'blank slate' imperative (the 'great reset' for post-COVID cities) has again been familiar in COVID responses, which emphasise quickness ('boldness') more so than equity and accountability (Badger 2020). The OECD summarised common city responses to COVID as: "digitalisation, mobility, density, urban design and collaborative governance and action-oriented guidance to build back better cities". Some are recognisable in Australia, and all carry both promise and risks.

Planning has a long and complicated relationship both with public health and with other crises and 'reset' moments, with a pattern of justifying and shaping planning to fit. Sanitation and health crises were used to justify and usher in a nascent planning profession and field of regulations in the early twentieth century. The rise of zoning involved complex interests aligning around a health-based crisis: with, in the US, courts validating the use of police power to zone land and stop the construction of a 'parasitic' apartment building is predicated on the quote "public health, safety, morals and general welfare". Health crises were also used as the basis for slum clearances in the mid-twentieth century. Oswald Barnett wrote of how "a heavy toll is being taken on the health of these occupants, particularly of the women and children", with this investigation leading to Victoria's Housing Commission and its city-shaping powers. The categorisation and problematisation of slums propelled rebuilding projects, themselves quickly (often) rendered as blights or failures needing further renewal.

Remodelling cities and planning in response to public health crises—then and now—is not altogether dissimilar to doing the same for economic and political ideas of crisis. Weaver (2017) traced the concept of specifically 'urban crisis' in the US and found that the term had since the mid-twentieth century been "mobilised to advance a range of political and economic interests". The urban crisis has been a term applied to a range of concerns: historically, to public health; to civil 'unrest' and upheaval; to austerity and fiscal crisis of the GFC; to inequality and urgent environmental pressures. The renewal work of Moses and others was openly grounded in a discourse of urban crisis: the diagnosis of crisis (and often—sickness) and the need for renewal justify absolutism. Scholarship on the governmentality of planning in a context of crises: fiscal, political, environmental and criminal (drugs) illustrates how many questionable renewal projects of the twentieth century were predicated on urban crises and how much fiscal austerity crises were used to justify the erosion of public processes.

Crisis, and a reshaping of planning to fit it, is also characteristic of neoliberalism. Corazza et al. (2021) argued that neoliberalism—as policy, as mindset and governmentality and as a material project—is not squished by crises (of economic crisis and austerity in Europe) but redeployed in a form of crisis response. It can remain "a coherent project" even when underscored by contradictory policymaking. Earlier, Hannah Arendt in The Human Condition wrote that the "economic miracle" of postwar Germany is a miracle only if seen in an outdated frame of reference". That rather, "under modern conditions the expropriation of people, the destruction of objects and the devastation of cities will turn out to be a radical stimulant for a process, not of mere recovery, but of quicker and more efficient accumulation of wealth" (Arendt 2013). The legacies of megalomaniacal urban renewal planners like Robert Moses and their positioning around 'urban crisis' share elements with the governmentality of economic and political crisis.

Austerity measures, too, are anti-crisis policies. Ponzini (2016) warned that in times of crisis "policymakers tend to lower the standards for regulation, to strip-off planning powers and authorities, to try to de-politicize and streamline choices and projects, to promote new strategies (e.g. smart city), in order to keep the available urban investments flowing and maintain political consensus among relevant stakeholders" (p. 1238). These are familiar patterns in Australia post-COVID. Planning has survived a series of crisis eras on account of its role in (partly in validating) neo-liberal society. Kunzmann (2016) called on planners not to just survive crises but to mobilise concerns about equity and social responsibility. They should raise their voice, he argued, to encourage debates based on disparities and equity issues. We should apply the same standards to a COVID response.

During COVID-19 public health experts unapologetically identified as experts were unambiguously referred to as experts and relied on technical experience and models to reinforce why a course of action was needed. There is often an explicit military tone to public health responses. (With a strange irony that the 'COVID Commander' in Victoria was formerly CEO of the state public transport agency.) This is in contrast to professional norms around planning. In Victoria, the premier would regularly announce that "I've listened to the experts", "I listen to the numbers". It is rare (perhaps unknown) that a planning minister would speak similarly of planning research. Perhaps planning researchers disagree too much or lack a clear enough imperative around things like car usage or design standards or climate mitigation. Nonetheless, researchers do put forward consistent evidence: many experts have been reinforcing the seriousness of man-made climate change. Housing affordability 'debate' and its disconnect with researchers and their emphasis on fairness is another case in point. That some fields of expertise are so routinely absent from urban politics and practice, while certain crises including COVID quickly recast what is politically possible is significant when considering the post-COVID city.

Public health and crises give momentum to urban projects, but also frame planning around human health and in a fairly narrow way. COVID responses embody a focus—managing density, reducing crowding, ventilation—on public health in which *urban* is dangerous and with it, people and their differences. Delay is also anathema to crisis responses. The softer urban planning terms for sudden, experimental changes include 'pop-up' and (in a more militaristic vein) 'guerrilla' or 'tactical' urbanism. David Webb referred to the effects of the economic crisis and its rollout of various forms of informal or 'tactical urbanism'. Sometimes these have been associated with progressive placemaking, but they have also involved complex interests and hybrid forms. So too with COVID responses, whether executive powers or pop-up car parks: a shift in the process can have beneficiaries and implications beyond the superficial and short term.

With crisis often comes a shift in governance and in the way government actions are justified. Naomi Klein referred to how crises—military, economic and environmental—have been used to usher in sudden 'shock doctrine' turns towards experimental neo-liberal policies. Since COVID she warned of 'screen new deal' technology beneficiaries. Klein quoted technology executives weighing in on New York's post-COVID reinvention: "we realise that change is not only imminent, but it can actually be a friend if done the right way", in "heading up a blue-ribbon commission to reimagine New York state's post-Covid reality". Their emphasis has been on rolling out ready solutions: as in "we need to look for solutions that can be presented now, and accelerated, and use technology to make things better", for example regarding education: "all these buildings, all these physical classrooms—why with all the technology you have"? Similar divisions and debates arose in Australia—the necessary or otherwise nature of previously non-negotiable community spaces. Amidst the fast-tracked infrastructure projects in Australia have been several, also fast-tracked multi-billion dollar commitments to digital precincts and digital governance.

In COVID and rapid responses to it, some of the obvious winners in Australian cities—aside from owners of beachside properties—have been technology and car companies. And the construction and infrastructure companies aligned with big build priorities. And while there have been unexpected resets and alignments around long-standing strategy including active transport and local centres—on the larger scale the winners from COVID city resets have also comprised technology companies and major retailers. Delivery and technology companies and the almost humorous image of what luckier Melbournians spent their lockdowns doing are powered by companies like Amazon and technology platforms. And have been delivered and materialised by supply chains composed of many often casualised gig-economy workers. Those who benefit from any sudden shift in cities and their governance are not going to be evenly distributed or necessarily progressive.

The COVID city has provided opportunities to reset modes of work, travel and daily interactions and has been embraced as an opportunity to reset public space and the regulations that shape it. COVID-related changes to cities have been much faster, and much more dramatic, than those either aspired to or attributable to planning policies in Australian cities. COVID-19 and the responses to it have fundamentally reshaped aspects of Australia's cities. We have shown that in some ways they have also reshaped or reset urban planning. We have also suggested that while the change in urban planning and transport has long been argued for in Australia, the kinds of sudden shifts associated with crises are fragile and to be taken on with caution. Jasiński (2022) contrasted two versions of responses to the post-pandemic city: on the one hand an emphasis on private space and suburbanisation, the return to the private car and a surveillance society. Or a possible positive urbanist and planning response: based on supporting a multifunctional and compact city, responsible and active transport and smart city tenets. In 'Streetfight' Jannette Sadik-Khan called for Moses-like action with Jacobs-like agendas. A challenge for Australian urban planning is to define an agenda—of equity, sustainability, evaluation, evidence and incrementalism—on its own terms rather than being reshaped by the higher profile crises of the day.

Our survey of planning change in Australia post-COVID indicates infrastructure projects are often front and centre. This is despite the possibility that while the migration is low, planners could in theory focus on making things work better and make more commitments to the practicalities of a 20-minute city or greening cities. In this window, planners could take an attitude of decisive action and act differently in key areas of problems around Australian cities. But planners should not assume that just because something dramatic is happening, and a 'great reset' occurring, 'good' change will happen without direct effort. It is also important to note that crisis narratives have short timelines: eventually, the 'fever pitch' fades ('living with COVID' replaces the daily breathless updates) and is replaced by long-term return to acceptance. In that later stage, the changes brought on the basis of the crisis do stay, for better or worse. Analogies might be found in traffic deaths, the long shadow of Australia's "6 o'clock swill" wartime measures on pubs that lasted until the 1960s, or any number of wars in the public televised consciousness. The crisis eventually becomes equilibrium. In the meantime taking ownership of long-term futures via a 'windows of opportunity' is, in the absence of more concerted attention by urban researchers and practitioners, as likely to be embraced by the urban plans of car companies, property interests and giant technology companies.

Some of the COVID health prescriptions for cities fit long-standing ideals, but in other parts are fundamentally anti-urban. They provide for different uses of public space, but also a city based on cars and private homes. The risk is of "throwing the baby out with the bathwater". At the same time, the COVID 'reset' idea exposes the flaws in the assumptions of the compact city argument in Australia, which posited that high densities were necessary for other forms of urbanism including public transport. We could now return to a more fundamental question—how do we live in a city where people have more access to green space—in a carbon-constrained world? Could this be done in a way that does not rely on private cars and on larger cities? Meanwhile, longer-term metropolitan planning strategies themselves do not seem to have been changed in Australia—either because of the effort involved, the actual lack of enforcement of many of these documents or because the underlying ideology of growth they channel is assumed to be irreversible. To argue that planning should not suddenly change or should not take direction from crisis-based technical experts: puts us in the unenviable position of defending a system that is unfair, ineffective and internally inconsistent. Planning *should* change. Possibly not in a pop-up 'great reset', 'fast track', crisis-framed way—again—because doing so is part of what needs to change.

References

Arendt H (2013) The human condition. University of Chicago Press

Badger E (2020) The pandemic has pushed aside city planning rules: but to whose benefit? The New York Times, July 20. https://www.nytimes.com/2020/07/20/upshot/pandemic-city-planning-inequality.html

Bliss L (2019) To build a better bus lane, just paint it. Bloomberg CityLab, March 1. https://www.bloomberg.com/news/articles/2019-03-01/howtactical-transit-lanes-help-buses-beat-traffic

Bolleter JA, Myers Z, Hooper P (2021) Overcoming the barriers to transit-oriented development. Aust Planner 57(3–4):188–205

Corazza MV, Moretti L, Forestieri G, Galiano G (2021) Chronicles from the new normal: urban planning, mobility and land-use management in the face of the COVID-19 crisis. Trans Res Interdisc Perspect 12:100503

Fowler M (2022) From pop-ups to permanent, bike lanes create new hurdles as city gradually refills. The Age, April 29. https://www.theage.com.au/national/victoria/from-pop-ups-to-permanent-bike-lanes-create-new-hurdles-as-city-gradually-refills-20220428-p5agrm.html
Freestone R (2000) Urban planning in a changing world: the twentieth century experience. Taylor & Francis
Gahbauer J, Matute J (2019) Best practices in implementing tactical transit lanes
Gratz R (2022) Robert Moses reconsidered: blight is in the eye of the beholder, City Limits. City Limits. https://citylimits.org/2007/04/02/robert-moses-reconsideredblight-is-in-the-eye-of-the-beholder/
Gurran N, Austin P, Whitehead C (2014) That sounds familiar! A decade of planning reform in Australia, England and New Zealand. Aust Planner 51(2):186–198
Gutman M (2008) Race, place, and play: Robert Moses and the WPA swimming pools in New York City. J Soc Archit Hist 67(4):532–561
Homes Victoria (2022) Big housing build. https://www.homes.vic.gov.au/big-housing-build
Hurley J, Taylor E, Dodson J (2017) Why has urban consolidation been so difficult. In: The Routledge handbook of Australian urban and regional planning, pp 123–135
Jasiński A (2022) COVID-19 pandemic is challenging some dogmas of modern urbanism. Cities 121:103498
Kunzmann KR (2016) Crisis and urban planning? A commentary. Eur Plan Stud 24(7):1313–1318
NSW Government (2020) COVID-19 recovery plan. https://www.nsw.gov.au/covid-19/covid-19-recovery-plan
Ponzini D (2016) Introduction: crisis and renewal of contemporary urban planning. Eur Plan Stud 24(7):1237–1245
Raynor K (2020) Victoria's $5.4 bn big housing build: it is big, but the social housing challenge is even bigger. The Conversation 18
Rowe H (2021) Is temporary the new permanent? COVID street experiments open our eyes to creating better cities. The Conversation
Taylor E (2020) Street fight: car parks, COVID-19 and the future of urban retail (part 1). https://www.foreground.com.au/planning-policy/street-fight-car-parks-covid-19-and-the-future-of-urban-retail-part-1/
Taylor E, Dalton T (2015) Keynes in the antipodes: the housing industry, first home owner grants and the global financial crisis. In: Housing in 21st-century Australia: people, practices and policies, pp 153–172
Victoria's Big Build (2022) Road and rail projects to transform how you travel, May 25. https://bigbuild.vic.gov.au/
WA Government (2022) COVID-19 planning reforms. https://www.wa.gov.au/government/document-collections/covid-19-planning-reforms
Wainwright O (2017) Street fighter: how Jane Jacobs saved New York from Bulldozer Bob. The Guardian
Wallock L (1991) The myth of the master builder: Robert Moses, New York, and the dynamics of metropolitan development since world war II. J Urban Hist 17(4):339–362
Waters C (2022) No more new bike lanes for CBD after council cops complaints, The Age, https://www.theage.com.au/politics/victoria/no-more-new-bike-lanes-for-cbd-after-council-cops-complaints-20220602-p5aqia.html
Weaver T (2017) Urban crisis: the genealogy of a concept. Urban Stud 54(9):2039–2055
Wynne R (2020) Planning revamp to boost economic recovery and jobs [Press release]. https://www.premier.vic.gov.au/planning-revamp-boost-economic-recovery-and-jobs

Chapter 8
The Future of COVID Cities

8.1 Are We There yet?

As we write this final chapter, the COVID-19 pandemic that has spread across the globe for the past two years is still having its impacts felt. Australia has come through the Omicron variant of the disease better than other countries, but not entirely unscathed. Our 'poster-child' status as being a country that successfully dealt with the initial outbreak of the disease was beginning to look a little tarnished. Having come to the last few months of 2021 being told there was light at the end of the tunnel, the government's 'let it rip' strategy meant, as one health worker in New South Wales put it, the light at the end of the tunnel might be a train!

The beginning of 2022 saw record cases, record deaths and for many people in Australia's cities, a return to some form of isolationist practices, even if not officially mandated. We started to see headlines like "People feel abandoned: as masks come off, thousands of Australians feel scared to go out" (Burke 2022). Many were again working from home (if in fact, they had actually gone back to the office). Some schooling was back online and events such as concerts and plays were cancelled or severely curtailed. Singing and dancing were once again banned. Health systems were put on high alert as more patients were admitted to Intensive Care Units. Victorian hospitals went into 'code brown alert', which just seems like an unfortunate colour choice to describe the situation they found themselves in. Case numbers increased so much that contract tracing, which had been a mainstay of public health management of previous outbreaks, was abandoned as the sheer task of tracking people's movements became impossible. So, despite promises of a summer reset, it seems that the wheels had indeed fallen off the country's attempt to live with the virus. Further into 2022, the country was still recording significant numbers of new infections, a problem compounded by the arrival of the annual flu season. People started to contract both COVID-19 and the flu (Flurona) at the same time.

All of this begs the question that if COVID-19 and its shape-shifting variants continue to hound us, what might be the long-term impacts on Australian cities and the residents who live, work and play within them? There are no guarantees that

S. Baum et al., *Pandemic Cities*, Cities Research Series,
https://doi.org/10.1007/978-981-19-5884-7_8

there will not be new more aggressive versions of the virus, nor that the vaccination protocols introduced will be effective in dealing with new variants. Two years ago, the idea that a mystery virus could shut down the world would have been considered far-fetched. The stuff of science fiction. But now, we wouldn't be surprised if a new variant popped up and started shutting things down all over again. Disappointed and worried, yes. But surprised, no.

In considering where we are heading, we should be mindful that cities throughout history have been shaped, in part, by responses to infectious diseases. As Lina Martinez and John Rennie Short have recently pointed out, "epidemics and pandemics played an important role in urban history" (Martínez and Short 2021, p. 2). Public Health concerns paved the way for the formalisation of parks and open spaces and cities were engineered to provide clean water, remove rubbish and treat waste, in part, to deal with urban areas as harmful places crowded with transmissible and deadly illnesses (Osborne 2008). As we saw in Chap. 3, it was John Snow's (1855) foray into health geography that showed us the public health links between the social structure of the city and the outbreaks of cholera and changed our understanding of how living conditions and hence city structure and form was associated with spikes in the cases of Cholera. Given our current situation, the question is how will COVID-19 and its lingering threat shape the cities we live in?

8.2 The Post-pandemic City

At the beginning of this book, we asked you to think back to early 2020 to the beginning of the pandemic. Now that we are on to our final chapter, we want to look forward. A question that has intrigued many of us has been what will the long-term impacts of the pandemic be on cities? Will cities transition into some kind of new normal or will they just transition back to pre-pandemic systems?

Let's be clear. It's obvious that cities themselves will continue to be a thing. After all, they have been around in some form for thousands of years and have survived threats to their existence before. Cities survived the post-World War I Spanish flu pandemic and cities around the globe have rebuilt and even reimagined themselves following civil skirmishes and natural or man-made disasters. There was once a time when increased global connectivity and the omnipresence of the Internet suggested that distances would no longer be an impediment to business, sounding a death knell to cities as we knew them. We all know now that that hasn't happened. For the contemporary city, there are significant economic forces at work that will ensure that business entities and individuals continue to cluster in meaningful agglomerations. Likewise, cities will likely continue to be the significant cultural and social nodes that have underpinned their vibrancy in the past.

But this doesn't mean we will not witness changes to the way we experience and navigate around our urban areas. There is likely to be some 'pandemic risk' (Glaeser 2020) especially as we come to realise that COVID-19 is unlikely to be the last globalised pandemic. Who remembers SARS, MERS, Ebola and H1N1? Sure, these didn't quite end up as huge global outbreaks, but they could have if given half a chance. And they all resulted in disruption to some extent. How the future pandemic risk is evaluated and how long the current pandemic keeps circulating will most certainly result in changes to our cities. While broad aggregate level patterns will likely remain unchanged, finer scale intra-urban or neighbourhood changes may take place. Some of the changes may have already been underway in some shape or form, but have simply been accelerated by the arrival of the pandemic and the public health and economic consequences that followed. Others may be venturing into new and untested territory.

Many of the changes to employment and shopping that were underway prior to the pandemic, but which have accelerated during the past couple of years, will result in lasting transitions. There is already significant evidence that employees are favouring, where possible, a split in working arrangements characterised by part-time spent in a formal office environment and part-time working from home. As we noted in Chap. 3, when asked about future working from home (#WFH) intentions, significant numbers of survey respondents confirmed that staying home for part of the week would be desirable. As #WFH becomes more normalised, we will likely see fewer people commuting the traditional five days per week into the office. With this, travel demand patterns and traditional peak hours may change leading to a transition in the ways we use our road networks and transport infrastructure.

The exponential rise in online shopping observed during the pandemic is likely to continue. While it was already a 'thing' prior to the emergence of COVID, shopping online exploded during the pandemic. Australian Bureau of Statistics data shows that online shopping rose 70.8 per cent in October 2020 compared to October 2019 (Australian Bureau of Statistics 2020). Large online merchants saw a many-fold increase in profits and contract delivery drivers dropping off parcels and gig-economy riders dropping off food became more commonplace in our suburbs.

These shifts may have implications for the form and function of our cities. Central Business Districts could see a shift away from the predominance of office and shopping hubs and morph towards something different. Geographer Richard Florida and his colleagues (Florida et al. 2021, p. 18) suggest that Central Business Districts may become significant cultural and civic hubs with "more events [taking] place outdoors, in city streets and plazas as part of the transformation of city centres into pedestrian and bike havens". In Australian CBDs, we have seen the emergence of 'pop-up' instalments of green spaces and bike lanes to help regenerate these forgotten urban spaces (Chap. 7).

The flip-side is that the pandemic has made many local neighbourhood centres more vibrant places. The requirement to stay in your local neighbourhoods or communities has revived many local areas. Some have argued that.

> With many more people working from home during the pandemic, there has been something of a retail inversion with more people shopping locally. There are clear signs of a resurgence in local shopping villages and high street retailing. There even appears to be a corner store revival of sorts (Mortimer et al. 2020).

Thinking locally may even extend to decisions around office space. Several commentators have noted the possibility of enterprises operating on a spoke and hub model of office location whereby a CBD office is connected to various smaller, de-centralised workspaces across the inner suburbs (Leggatt 2021). This, in turn, will see a rethink of central office space requirements and a concurrent (re) vitalisation of neighbourhood centres.

Post-pandemic cities will regain their vibrancy in some form or another. As international borders open and tourists and students return, tourist areas of our cities will begin to buzz once again. Inner-city university campuses, which witnessed a decline due to the lock-out of international students (Chap. 2), are coming back to life. Combined with the desire for most Australians to get on living their lives, busy weekend socialising precincts are once again getting that crowded feel about them.

The post-pandemic city will almost certainly maintain its unequal socio-economic nature. The patterns we discussed in Chap. 2 have been a feature of our cities for generations. And sadly, it is often the same neighbourhoods and suburbs that are the most disadvantaged time and time again. To steal a line from the movie Casablanca, 'round up the usual suspects'. The COVID pandemic and the public health and economic response have only reinforced these unequal patterns. In many cases, it will have made things worse. As some have suggested, "the wealthy face minor inconvenience, but the poorer face economic collapse" (Martínez and Short 2021, p. 6). We know from previous recessionary periods and from analysis undertaken during the pandemic (Baum 2021) that low-income and disadvantaged regions and communities take longer to recover from a downturn than more advantaged places. The social problems associated with concentrated disadvantage across our cities will be further entrenched due to the unintended but likely intergenerational effects imposed by the government's public health measures. School kids in overcrowded houses with poor Internet access trying to home school don't have the same life chances as those from more affluent places, thereby further exacerbating existing problems.

One important question that still needs to be played out is what, if any, impact will COVID-19 have on the way we think about urban areas and their design? Will we see greater decentralisation of office space away from centres to outer-suburban regions? We saw during the pandemic that the use of outdoor green space was at a premium and that some areas were better served than others.

This begs the question will we see a greater demand for new green space developments? Then there is building design. We have learned that some building forms, especially in residential spaces, are poorly set up to deal with the requirements of self-isolation and multiple work at home requirements. Designers in the United States are already introducing housing designs to better meet the needs of the pandemic world. One demonstration home looks just like any normal four-bedroom house, except it includes.

> A front-door "vestibule" for no-contact home delivery, two "pocket" home offices (neither of which is a bedroom), a school room, a quarantine room, a "secret room" behind a bookcase (for being alone), and a whopping *four* bathrooms (Purtill 2022) (para 3).

At the very least, people will be demanding more space and greater flexibility of use, so as to allow living spaces to be more efficiently utilised.

8.3 What Have We Learned

Close scrutiny will show that most crisis situations are opportunities to either advance or stay where you are (Maltz 1978)

Given that we are now two-plus years on from the beginning of the pandemic, it is useful to begin to consider what we have learned. If we consider that a crisis is an opportunity to advance as Maxwell Maltz suggests, then what should we be looking out for? What we have witnessed over the past two years is reflective of a range of dynamics that were in place prior to the outbreak. The patterns, consequences and outcomes we have written about in the preceding chapters underline the multitude of issues facing governments, policymakers, academics and the communities we all live in. At the same time, they offer useful paths forward, if we choose to listen and investigate the possibilities.

There are clearly going to be disagreements about what are the most important take-home messages from the pandemic, but here are a few. Importantly, the lessons apply not only to the pandemic and its impact on cities and communities. They can equally be thought of as applying to any crisis. The obvious example is climate change but think about other man-made or natural crises as well.

As the pandemic expanded across cities and communities, it exposed the need for good governance and leadership, something that those living through previous crises such as the summer bushfires and floods had failed to witness. The pandemic had governments acting in an environment of radical uncertainty. COVID was a 'wicked problem' (see below) that required multi-agency, multi-level cooperation and input. It was all at once a health, economic, security and migration issue. And it was an issue that crossed all levels of government from local and state to federal. The pandemic showed us what could be achieved if governance structures were good and functioned well. It also demonstrated what happens when things slip.

The federal government's fiscal interventions showed that governments could unshackle themselves from their neo-liberal constraints in order to support the most vulnerable. Although these interventions did not go far enough, either in scope or level of support, they did show what is possible when politicians from all sides of the political divide "leave their ideologies at the door". At no stage during the early rollout was there the usual hand-wringing about where the money was going to come from. For once the decisions were being driven by the politics of helping (some) people rather than the politics of budget repair, austerity and small government which have resulted in decades of reduced services and the ability to properly respond to the crisis in the first place. In cities, but also elsewhere, the pandemic showed what happens when governments cut back on health spending and privatise essential 'public good' services, in the name of efficiency.

But the pandemic also showed what could be achieved if multi-levels of governments worked together. The announcement of the national cabinet in early 2020 which involved the federal government working with their state counterparts provided some promise. Our form of government and the different responsibilities held by different levels of government meant that a unified approach involving all stakeholders was imperative. While it lacked cohesive leadership and often seemed to descend into a battle between the states and the federal governments, it did provide a glimpse into what was required in such a crisis and potentially the good work that could be achieved.

There is no doubt that the problems faced by those who were charged with guiding Australia through the crisis were complex, and that this complexity meant that perhaps unlike other crises (except perhaps climate change) leadership was ill-equipped to deal with the problems as they rapidly came forward. The arrival of the pandemic introduced a range of interconnected health, social and economic problems. It was by many definitions a 'wicked problem' characterised by reduced time windows within which action can be taken, the lack of a clear stopping rule, the unknowns about success and failure, the complex interplay between individual and collective networks and the existence of near-term policy approaches focussing on shorter-term economic and interest-based calculations rather than future-focussed policy ambitions.

The need to deal with the pandemic in short time frames and without complete knowledge resulted in a range of unintended consequences, especially when policies and directives were seen as being made on the run. Knee-jerk actions such as the early shutting down of entire blocks of public flats in Melbourne (see Chap. 1) were considered by many as early examples of policymaking on the run without complete information. Such was the character of the rapidly changing pandemic.

Some of the unintended consequences might be viewed as positive gains—the increase in active travel (Chap. 6) or the wider acceptance of work flexibility (Chap. 4)—but, there was also a range of negative impacts. The rapid rollouts of some fiscal support measures by governments, for example, were hit and miss, with the government seemingly changing the framework on a political whim. Some people were included and some were excluded. The first tranche of the government's Jobkeeper payments meant that in theory at least, some casual workers would receive a windfall (this was later changed). But many were excluded. International students missed out. Public universities were excluded. Casual workers who hadn't been in employment for 12 months were ineligible. Local government employees were excluded. So, although the government's Jobkeeper program aided many people, it also resulted in increasing hardship for many others. We soon learned we were not all in this together. The pandemic brought into plain view the layered social and economic landscapes and underlying inequities at play in our cities. For the most disadvantaged, the unintended consequences of the pandemic often hit hard, further exacerbating existing problems.

Stay-at-home orders meant that people were actively staying away from large social and work gatherings where the risk of the spread of the virus was greatest. But the consequences of forcing people to isolate themselves at home ranged from the minor annoyance of home schooling to more serious issues around increased domestic violence cases. During lockdowns, domestic violence service providers reported increases in demand and reported an increase in controlling and coercive behaviour (Hermant 2021). There were also knock-on effects on routine health screening as people in lockdown were unwilling to venture out to the doctor. And there were the unintended impacts on a range of activities that in normal times would not be subject to restrictions, but which COVID brought to a screaming halt. Think about children's sporting clubs, social groups, Saturday golf competitions, church services, picnics and long lunches, all of which provide communal glue and facilitate individual well-being disappeared almost overnight (Carroll 2020).

The complex nature of the pandemic also had significance for the process of crisis communications. The need to get information out quickly, combined with the range of views about the virus, its origin and its future meant that crisis communication often lacked cut-through. Those with poor language skills often missed out on important messages, while a lack of clear scientific messages muddied by misinformation meant that Coronavirus scepticism was sometimes rife.

Perhaps it was a concern for the lack of clear and unified leadership or because of the sense of helplessness many people felt, but throughout the whole confusing mess, community, in all its forms, became much more important. The disruption of stay-at-home orders meant that in many places, local communities became necessarily more connected and, in turn, vibrant. The physical form of communities within cities, those places where people were riding out the pandemic lockdowns, became interconnected places of care and service. There is evidence that neighbours helped each other out with food runs and the provision of other services, especially when an individual or household had to self-isolate. One study suggested that 40 per cent of Australians thought their relationship with their neighbours had transformed since COVID-19, with 42 per cent talking to their neighbours more and 63 per cent much happier speaking with their neighbours more often (Suncorp 2021). Where contemporary cities have been criticised as often being the death of neighbouring and the sense of community, it seems that the pandemic may have altered this, although for how long no one can be certain. The famous "mud-army" that swung into action following the widespread 2011 floods in Brisbane and South-East Queensland saw communities and neighbours pitch-in to help in the clean-up, although the community spirit eventually waned after the immediacy declined.

Alongside "real" bricks-and-mortar communities, virtual communities also became far more important. We have seen in the past that disasters, crises and emergencies have motivated people to share information, human contact, emotional support and even just basic conversation in virtual online forums (Al-Omoush et al. 2021). COVID-19 saw an increase in online traffic in virtual communities on platforms such as Facebook as people got online to share news, and concerns, and ask for help.

8.4 Where to from Here?

Where to from here? It is clear that Australian cities and the people and communities within them are not yet over the hangover introduced when the global pandemic burst onto the scene in early 2020. But the pandemic and its management have taught us a lot about how we should deal with crises, what works and what doesn't, who are the most vulnerable and how crises are not necessarily the catalyst of problems but simply act to magnify the one already there. Going forward we can only hope that for all the anguish, disruption and turmoil the pandemic brought to our cities, we all learn something useful. Hopefully, this book at least starts some fruitful conversations.

References

Al-Omoush KS, Orero-Blat M, Ribeiro-Soriano D (2021) The role of sense of community in harnessing the wisdom of crowds and creating collaborative knowledge during the COVID-19 pandemic. J Bus Res 132:765–774

Australian Bureau of Statistics (2020) Online sales, October 2020—supplementary COVID-19 analysis. Retrieved from https://www.abs.gov.au/articles/online-sales-october-2020-supplementary-covid-19-analysis

Baum S (2021) Using a regional lens reveals the uneven impact of the COVID employment crash. Retrieved from http://bilbo.economicoutlook.net/blog/?p=46861

Burke K (2022) People feel abandoned': as masks come off, thousands of Australians feel scared to go out. The Guardian. Retrieved from https://www.theguardian.com/culture/2022/apr/14/people-feel-abandoned-as-masks-come-off-thousands-of-australians-feel-scared-to-go-out

Carroll J (2020) Coronavirus: Shakespeare's dreamlike vision could become our sick, dystopian nightmare. The Australian. Retrieved from https://www.theaustralian.com.au/inquirer/coronavirus-shakespeares-dreamlike-vision-could-become-our-sick-dystopian-nightmare/news-story/73ca636f3d0902cd1ad99f11c4640bb1

Florida R, Rodríguez-Pose A, Storper M (2021) Cities in a post-COVID world. Urban Stud 00420980211018072

Glaeser E (2020) Cities and pandemics have a long history. City J

Hermant N (2021) Domestic violence surging amid COVID-19 lockdowns, research shows. ABC news. Retrieved from https://www.abc.net.au/news/2021-06-25/coronavirus-covid-lockdowns-and-domestic-violence-data/100237406

Leggatt J (2021) The future of cities in a post-COVID world. In: The black. Retrieved from https://www.intheblack.com/articles/2021/04/01/future-cities-post-covid-world

Maltz M (1978) Conquest of frustration. Ballantine Books

Martínez L, Short JR (2021) The pandemic city: Urban issues in the time of COVID-19. Sustainability 13(6):3295

Mortimer G, Grimmer L, Maginn P (2020) The suburbs are the future of post-COVID retail. The conversation. Retrieved from https://theconversation.com/the-suburbs-are-the-future-of-post-covid-retail-148802

Osborne J (2008) Preparing for the pandemic: city boards of health and the arrival of cholera in Montreal, New York, and Philadelphia in 1832. Urban History Rev/revue D'histoire Urbaine 36(2):29–42

Purtill J (2022) Four bedrooms ... four bathrooms? How COVID-19 is changing home design. *ABC News*. Retrieved from https://www.abc.net.au/news/science/2022-05-20/covid-19-pandemic-home-house-design-how-it-is-changing/101077024

Snow, J. (1855). *On the mode of communication of cholera*: John Churchill.

Suncorp (2021) Love thy neighbour: Almost half of Aussies now closer with neighbours than before COVID-19. Retrieved from https://www.suncorpgroup.com.au/news/community/aussies-now-closer-with-neighbours